The Evolution of Speech and Language

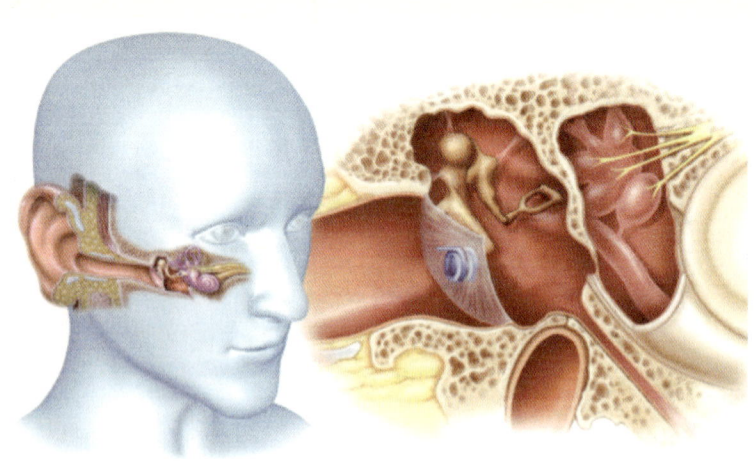

The Evolution of Speech and Language

DN Lele MS (ENT), DLO, FICS

Director
Lele Hospital and Research Centre
Nashik, Maharashtra

CBS

CBS Publishers & Distributors Pvt Ltd

New Delhi • Bengaluru • Chennai • Kochi • Kolkata • Mumbai
Hyderabad • Nagpur • Patna • Pune • Vijayawada

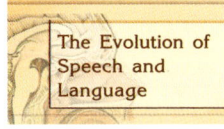

The Evolution of Speech and Language

ISBN: 978-81-239-2956-9

Copyright © Author and Publisher

First Edition: 2016

Published by Satish Kumar Jain and produced by Varun Jain for

CBS Publishers & Distributors Pvt Ltd
4819/XI Prahlad Street, 24 Ansari Road, Daryaganj, New Delhi 110 002, India.
Ph: 23289259, 23266861, 23266867 Website: www.cbspd.com
Fax: 011-23243014   e-mail: delhi@cbspd.com; cbspubs@airtelmail.in.
Corporate Office: 204 FIE, Industrial Area, Patparganj, Delhi 110 092
Ph: 4934 4934   Fax: 4934 4935   e-mail: publishing@cbspd.com; publicity@cbspd.com

Branches

• **Bengaluru:** Seema House 2975, 17th Cross, K.R. Road,
  Banasankari 2nd Stage, Bengaluru 560 070, Karnataka
  Ph: +91-80-26771678/79   Fax: +91-80-26771680   e-mail: bangalore@cbspd.com
• **Chennai:** 7, Subbaraya Street, Shenoy Nagar, Chennai 600 030, Tamil Nadu
  Ph: +91-44-26680620, 26681266   Fax: +91-44-42032115   e-mail: chennai@cbspd.com
• **Kochi:** Ashana House, No. 39/1904, AM Thomas Road, Valanjambalam,
  Ernakulam 682 018, Kochi, Kerala
  Ph: +91-484-4059061-65   Fax: +91-484-4059065   e-mail: kochi@cbspd.com
• **Kolkata:** 6/B, Ground Floor, Rameswar Shaw Road, Kolkata-700 014, West Bengal
  Ph: +91-33-22891126, 22891127, 22891128   e-mail: kolkata@cbspd.com
• **Mumbai:** 83-C, Dr E Moses Road, Worli, Mumbai-400018, Maharashtra
  Ph: +91-22-24902340/41   Fax: +91-22-24902342   e-mail: mumbai@cbspd.com

Representatives

• **Hyderabad** 0-9885175004 • **Nagpur**   0-9021734563 • **Patna** 0-9334159340
• **Pune**   0-9623451994 • **Vijayawada** 0-9000660880

Printed at: HT Media Ltd., Noida

to

---

many infants I came across last many years
who were deprived of speech. Result of hearing loss and
Principal, Aaarati Khaladkar and a team of
her dedicated teachers who did their best to give
them speech and bring them to mainstream
education and acquire many academic qualifications

Foreword

Language has been with humans from the very beginning of pre-history. It will not be any exaggeration that it was language gaining of the language ability that differentiated humans as a species from the pre-human ancestors. Today, we cannot think of language without thinking of it in the form of writing or various other representations of speech; but, the writing stage entered human civilizations relatively far more recently than the ability to use speech for transaction of emotions and meaning. The human journey on the path of discovering language as a means of communication has been painfully slow. The discovery of speech therefore cannot be easily compared with the discovery of fire or tools. Probably, it can be compared with man's realization that the peculiar placement of the thumb allowed human to grapple with the environment more flexibly than an animal paw would do. But even that pales in comparison when we notice that the time required for moving from making the first guttural and nasal sound to producing the first full sentence has been nearly 400 thousand years. Compared to that prolonged phase of acquiring the language ability the period of the last seventy thousand years during which we have been using language has been relatively shorter. Yet, during these millennia, language has come to occupy such a centrality in the mental and practical transactions of humans that it would be nearly impossible to think of humans without language as a 'form of life'. It has been the most important means of man's negotiation between the consciousness and the world that stretches extensively outside us on our planet and beyond it. It has also been the primary means of storing memory and thereby conceptualizing time as well as managing the relationship between the present and the past. Space as we understand it is 'represented' in language before it becomes intelligible for us, time is retrieved in and through language alone. Yet, despite being so crucially important for the functioning of the human consciousness and intellect, language as a phenomenon has still remained surrounded in several mysteries regarding its source, its exact functioning in the neurological space, its psychic appeal and its ability to represent and interpret hugely complex shades of ever-shifting meaning. Considering the fundamentally crucial association of language with the very evolution of the human species and also considering how complex a phenomenon. It has been as a material system as well as a mental flux. It is not at all surprising that language has remained through the ages as the most prominent among the subjects of philosophical enquiry and scientific exploration. The Avesta and the Vedas ascribe the creation of the world to the creation of word. Philosophers in all civilizations have repeatedly tried to grapple with the mystery that language is. Grammarians have tried to present classificatory descriptions of meaning is conveyed, mystics have dwelt upon the magical powers it has and poets have wooed it for seeking its evocative charms. In our time, language has returned to lure the scientific enquiry by its ability to provide technological breakthroughs in the form of universal translation and fascinating communication tools never imagined before. However, in the scientific pursuit of language, India seems to lagged behind during the last couple of centuries despite our producing the mesmerizing literary diversity that makes the Indian people and thought what they are? When one thinks of India's contribution to enhancing the human understanding of language, one normally turns back to Panini of the pre-Christian era or Bhartrihari of the third century, or at best to the *dhvani* and *sphota* theories of the classical age. But, for the recent past, we have, unfortunately, a little to show. There have, of course, been a number of noteworthy Indian scholars of the nineteenth century and the twentieth century whose descriptive and interpretative studies have been noticed world over as highly original. But, at the beginning of the twenty-first century, if an Indian scholar has to think of any fundamental issue related to language—whether of structure, of significance, or its genealogy or its evolution—one cannot help but having to refer to the thinkers and authorities of the west, Herder, Schiller, Saussure, Wittgenstein, Derrida, Chomsky and such. In other words, India has not much to show by way of any really fundamentally radical departure or new realization in thinking about the language phenomenon. I need not say how elated one may feel if, in these circumstances, one suddenly realized that a lonely but dedicated thinker or scientist has managed to arrive entirely by himself at a major breakthrough? I had this kind of feeling when I was first told about Dr Lele, a medical doctor working from an academic outpost as the city of Nasik in north-west Maharashtra. I was told about him by my friend Dr Kumar Ketkar, a distinguished editor of several newspapers. When Dr Ketkar told me about Dr Lele, my first and-now I know-naïve impression was that he might have collected some case studies of speech disorder or malfunctioning. But when I spoke to over telephone, he told me the passion he had put behind his lifelong pursuit of understanding how and when humans developed the physiological basis for producing speech? I must confess that I was completely bowled over by the statement of the ambitious project. Then, soon after, Dr Lele decided to send his working manuscript to me for my opinion. As I stared reading it, I found it unbelievably original and perceptive. It was not a work of a person obsessed with the originality and validity of his own views. It was rather an extremely well-calibrated scientific treatise, written as if he would be writing a scientific textbook for the advanced postgraduates in a branch of medicine, with evidence and reasoning explicated at every stage. What impressed me the most about Dr Lele's work was that while he knew that he was proposing an original thesis, sure to be noticed and discussed in learned circles, there was nothing precocious about him reflecting in his style? His work that was done in the service of enhancing the human understanding of the history of man's acquisition of the language ability, presented in the most respectable tradition of scientific inquiry, simple, clear and convincing. It must at once be stated that Dr Lele's work was not easy. He had to go over a vast range of literature dealing with various aspects of the evolutionary biosciences, phonetics, semantics, neurology, and linguistics. What is remarkable about his study is that in presenting his own findings and interpretations, nowhere in the work, Dr Lele is tendentious. He has followed a scientific objectivity in arriving at his own conclusions. The conclusions are in his own words.

The sound wave traveling through the ear, cochlea, getting converted in an electrical wave reaches brainstem. From there it travels to the auditory center in left temporal lobe, then reaches Wernicke's area and is comprehended, and is then either answered or stored in memory. It is its last station. It is logical therefore when you want to speak, a new wave is generated in the Wernicke's area only. Wernicke and angular gyrus are parts of IPL and here the auditory wave is translated or adapted parallel to the visual profile of object or action received via visual sensation received by angular gyrus. This takes place in presence of mirror neurons. From here auditory wave reaches the Broca's area. Neural circuits are already present there for gestures since millions of years. They split and by exapatiation of function render auditory wave exactly matching to the visual profile of the object or action, in the presence of mirror neurons. It is a cultural origin or evolution making use of structures already evolved for some other purpose or are still performing those purposes. In this theory there was no need of taking help of any innate factor such as, language organ, universal grammar or language acquisition device (LAD) or even an abstract mental space. Furthermore a highly complex set of behavioral or cultural system like human language are likely to be influenced by many genes not by only one. Mutations will typically be observable only rarely when their effects are unusually large and disrupt many systems. This is the complete story of origin of our faculty of speech and language. This is the special attribute that distinguishes modern human from all other species existing or existed before on the globe.

The hypothesis forwarded by Dr Lele and the fascinating range of proofs and evidence provided by him to forward his theory of how the evolutionary process has multi-tasked the neurological and physiological apparatus provided within the brain and the body for making language possible as a cultural product, deserve to be entirely welcomed and calls for an engaged discussion on the intimate link between culture and nature, between human triumphs and the process of natural selection? A serious thesis has been presented by Dr Lele on the process of synesthesia and cognition, on the inter-domain neurological powers man as a basis of creating as complex a cultural phenomenon as speech-based communication.

I whole heartedly welcome the work as a seminal work by a contemporary and dedicated Indian. It is no ordinary intellectual achievement.

GN Devy
Founder
Bhasha Research Centre, Baroda

Foreword

I consider it a matter of immense privilege to write the foreword to this book by the very revered and sagacious Dr DN Lele. As the great philosopher Immanuel Kant says—

"All our knowledge begins with the senses, proceeds then to understanding, and ends with reason, there is nothing higher than reason".

It is the quest for finding out this underlying "reason" that makes a man extraordinary. I feel it is this heartfelt quest "to find a reason" for the linguistic origin that has made this wonderful treatise on *The Evolution of Speech and Language*. The evolution of language is a very integral subject to evolution of mankind. However, as per my knowledge, not much has been written on this. Also this subject has a multidisciplinary approach, with contributions expected from neuroscientists, otopharyngologists, anatomists, anthropologists, linguists, etc. Such a broad-based work has been missing in such an important area of linguistic development. However, this book fulfills this gap by providing a relevant piece of information from all these sciences and more, and placing them together like a jigsaw puzzle to make a complete comprehensive picture.

Dr Lele starts from a very basic question—"What is the difference between language and communication"?

Interesting is not it?

He takes us through a detailed relevant history of evolution of man to find an answer to how and when exactly did the *Homo sapiens* acquired this art of speech?

It is visibly evident that Dr Lele has taken enormous efforts to explore the evolution of language and speech from all possible aspects. From Charles Darwin's theory of natural selection to Dr VS Ramachandran's Baoba-Kiki experiments he traces the change in perceptions and scientific beliefs about evolution of speech.

The metaphysics of origin of language is beautifully explained in the Hindu philosophy through Kashmir Shaivism where speech is considered to exist at multiple levels. The exterior (or spoken) speech is expressed through the 'vaak tattva.' The full scale speech is expressed through sub-phases: Parra, pre-speech, 'Mdhyama vaak'—mental speech, interior 'Vaikhari vaak' spoken word, exterior.

Dr Lele has described the noteworthy contributions of ancient Indian scholars Panini and Katyayan in the development of the language. The correlation of ancient Indian wisdom with modern findings certainly deserves appreciation.

My specific interest was aroused while reading the chapter on Mirror Neurons, as I am myself is involved in writing a book on Mirror Neurons. He has beautifully explained how Mirror Neurons may form the link between visual sensation and auditory inputs to understand the neurological basis of development of speech and language?

The most scoring highlight of the book is that Dr Lele has proposed his own hypothesis of evolution of speech and language after an sightful consideration of all the major prevalent theories by other scientists and workers in this field. He does not merely proposes a theory based on futile assumptions, but conceives a theory about an unseen phenomenon of origin of speech, then substantiate this theory with rational reasoning from numerous scientific evidences and then ultimately reconfirms his theory on the grounds of these sturdy evidences.

To me this book though the title may dubiously appear scientific, is made for readers from all kinds of backgrounds, but with an open mind, to grasp the essence of our existence as a human and what makes us so 'human'. Dr Lele has a very simple yet gripping way of writing. The book is strewn with numerous short stories, anecdots, and contemporary references. It is almost like a grandfather telling tales of wisdom and discoveries to the young generation, through his years of experiences.

At this ripe age, going the extra mile with writing this book is an act of pure selflessness ...... that is to gift a legacy of knowledge to the future generations interested in this area.

I once again congratulate Dr DN Lele for giving us this very much needed treatise in the field of linguistic origin and development. I am sure the readers will be glad to experience the intellectual joy of knowing what it really means to be able to communicate through "speech" and understand the true power of this exclusive gift of the Lord Almighty to the mankind.

Dr Arun Jamkar
Vice-Chancellor
Maharashtra University Health Sciences
Nashik

Prologue

I AM NOT A LINGUIST

I am an otolaryngologist by profession. What inspired me to write on a subject like "Evolution of Speech and Language", an origin of linguistics itself?

In the last 20 to 30 years, many linguists and neurologists have tried to explain the origin of speech and language with numerous theories. Famous neurologist of Indian origin Dr Ramachandran has written an article "The power of babble" in his book 'The Tell-Tale Brain' about his concepts of origin of language. To the best of my knowledge no otolaryngologist, or so to say any one else in the medical field has thought it worthy of writing on this subject. A new branch of 'phonosurgery' is upcoming in the field of laryngology. Actually this subject should belong to the field of basic sciences in this field. Even then no one belonging to the field has thought of going to the root of it. As an ENT surgeon that was one reason why I thought that it was my field and why not try myself thinking about it.

Is that the only reason why I decided to take up this subject? The real urge for me to take up this subject was different. The story starts about 50 years ago. In 1964–65, I had been to UK to learn, stapes surgery for otosclerosis and tympanoplasty surgery for conductive hearing loss, secondary to discharging ear. Otosclerosis is a disease of young people and specially affects the young women during pregnancy. Formerly the surgery for this, was very extensive with a big skin incision outside and a hospital stay for about seven days, and included occasional side effects. The patients therefore had a fear for it. In 1961, John Shea of states, introduced a new technique by which the surgery became very easy. There was no need of external incision, practically no hospital stay and no side effects. This was a miracle in the field and I very much wanted to learn it. Small story will be very interesting.

A young girl of about 18 was brought to me for bilateral conductive hearing loss. She was coming from a long distance, from another state with a great faith. Her marriage was fixed and the parents did not want the other party to know about her hearing loss. The surgery went smooth and the marriage was celebrated with all the enthusiasm. Happy ending was, her husband brought her back after a few months with all the confidence, for the surgery of the other ear. The couple went back very, happily and a basketful of thanks to me.

While in UK working in a hospital I came across an audiogram (graphic record of hearing loss) of a child of about 2 and ½ years old. Till that time there was no test in our country for testing such a small kid for hearing loss. Out of curiosity I visited the center where such tests were undertaken. There I learnt about the special schools for training these children. They were not only, having hearing loss but also had not developed speech because of their hearing loss. In the schools they were not only learnt to hear but they were given lessons to speak also. There and then only I took a decision to start such an institution on my return back. The only difficulty was having an assistant and a teacher for hearing impaired. My better half Shrimati Vaijayanti who had visited UK took interest in visiting these institutions and tried to grasp the technique as much as possible during her short stay of 2 months. When she returned, after some time she obtained a regular diploma in training and teaching hearing impaired. At about the same time a great freedom worker of Nashik Shrimati Kusumtai Patwardhan and her Husband with their friends had started an educational institution. They welcome our idea of starting an institution for hearing impaired in Nashik. The very thing that in a short span of 35 years the institution has spread with many departments was enough to prove the importance of this program.

My Interest in Linguistics

We had started a diploma college for teachers of hearing impaired. Speech and language training was a part of the program. There were lecturers for different subjects, but I decided to learn something more about speech and language as these were the part of my specialty. I therefore started reading something basic about language and linguistics.

One more reason for this curiosity was, these children are brought to an otologist for non-development of speech, parents rarely know that the reason for non-development of speech was really the hearing impairment of the child is hearing impaired, as it has never heard what to talk. The hearing impairedness is an **invisible disability.** As these children are without speech they have to be taught both to hear and talk.

I therefore thought that I must have some basic knowledge about language and linguistics. I learnt about our heritage names like Panini, Patanjali, and Katyayan who had done a lot of basic work in this field and written treatises. I also read a lot from contemporary authorities and their work from developed world.

Language gives us opportunity or a medium to exchange our thoughts and enrich our cultural and scientific development. It has helped us to achieve spectacular developments in all the branches of science, so much so that man has stepped on the moon and is now thinking of keeping his foot on Mars. Language is a special attribute of man. The father of theory of evolution Darwin has said neither Chimpanzee nor any pre-humen species had speech. They were communicating with vocal massages. How then *Homo sapiens* acquired this art of speech and when?

When I decided to write about "Evolution of Speech and Language" I realized that the subject touches many disciplines of science. First I will have to go down a few thousands or perhaps millions of years to trace the history of mankind. We (humans) belong to vertebrates therefore I will have to go to the development of and comparative study of anatomy and physiology of the related systems, how it changes in human to get the art of speech? And therefore the subject touches, archeology, anthropology, evolutionary biology, comparative anatomy, neurology, neurosciences, electronics, computer sciences, and even music. No one person, can be expert in all these fields. Still I had to go through the part relevant to my present subject, from all.

In last 50 years or so many new investigative techniques have evolved and become available to scientific field. Some of them are cineradiography, PET, FMRI, electroencephalography, electromagnetography and many others. We will learn more about them at the end of this section. Most important factor is these tests have helped us to learn more about live functioning of brain, articulatory system and many other physiological systems. Since then many scientists, belonging to many branches of science, mainly linguists but conspicuously barring medicine, have started writing about "The evolution of speech and language" none was unfortunately giving a right answer to this puzzle and it remained a puzzle indeed.

When I went to meet Padmavibhushan Dr Ashok Kelkar, the world renowned linguist, to tell him about my intention of trying to solve this puzzle, he was very happy and welcome the idea and encouraged me to go ahead. He said "The evolution of speech and language" is the most neglected part of linguistics, perhaps it is a stepchild of linguistics. The reason he gave was, what I have stated above, it is a multidisciplinary subject and is having a dominant medical background? Perhaps what he wanted to say was thinkers from either side are siting back to back? The caricature accompanying this will give the exact idea of the exact situation.

To train the teachers for handicap it was necessary to give them training in speech and language. There were lecturers for the subjects, but I thought of learning myself something more about it. I therefore started reading something basic about language and linguistics. One more reason for this curiosity was, these children are brought, to an otologist for non-development of speech. Parents never accepted that the child can be deaf as deafness is an **invisible disability.** Language is a special attribute of man. The father of theory of evolution Darwin has said that Chimpanzee and pre-human species had

no speech, and they were communicating via vocal messages. How then did we, the *Homo sapiens* acquired this art of speech and language?

A new branch of phonosurgery is upcoming in last few years in the field of laryngology. Naturally the subject of the evolution of speech should belong to the field of basic sciences in this field. As an ENT surgeon that was one more reason why I thought that it is my field, to write about it.

When I decided to write about evolution of speech I realized that the subject touches many disciplines of science. Initially I will have to go down a few thousands of years or perhaps millions of years back, to trace the history of modern human. This subject encompasses many other related subjects such as anatomy and physiology of related systems, archeology, anthropology, evolutionary biology, comparative anatomy, neurology and neurosciences, electronics, computer sciences and even music. No one, can be expert in all these fields, but still I had to go through all these disciplines, at least the part relevant to my present subject. Present volume is the result of these efforts. Attempt is made to make the subject easy to understand by ample figures and charts. Moreover we will be taking a short review of what we will be reading in each chapter that will help the reader to go through the subject matter easily.

The first chapter will explain the difference between the communication and speech. Animals and pre-humen species were also communicating with each other, but with the help of means other than speech. P Lieberman in his book "Eves Spoke", has said that, animals who talk are humans, because what sets us apart from other animals is, we have is the "Gift of speech?" We will have a historical perceptive from mythology to medieval period. The work of many authorities in that period and about their concepts of the puzzle of 'The origin of speech and language'.

If human is the only animal having speech and compositional language **it becomes essential to think as to what can be the cause of it.** Is there any basic difference in the anatomy, and physiology in the structures concerned with this? This will take us to the evolutionary biology and process of evolution of modern man. We will go through this in brief in the second chapter.

If we wish to study 'The evolution of speech and language'. we will first have to know what is the basic difference between the vocal communication of animals and speech of modern man? How a modern man speaks? All this will be explained in detail in Chapter 3.

Language is not a single package. It can be expressed in statements or can also be expressed in musical form. Music is not certainly an auditory cheesecake as Pinker would have us believe. How was music originated? Did it come with language as an evolutionary spin-off? Or a lucky break? Music provides a relief from tedium of survival—the basic instinct and therefore is it deeply rooted in our biology? Is it encoded in the human genome during the evolutionary history? Where the center for appreciation of music is located in the brain? What are the benefits of music itself, and when it comes accompanied with a lyric? All this forms a part of study, when we think of origin of speech and language. This is the purpose of including the fourth chapter in this treatise.

When one says, that the human is the only animal having speech it becomes imperative for us to prove that no other species existing today other than man or which existed before had speech. The species immediately previous to our species— *Homo sapiens* in the process of evolution, is *Homo neanderthalesis*. They evolved from the same species *Homo heidelbergensis* in Europe, from which we evolved, though in Africa. Neanderthals evolved 300,000 thousand of years back. That is about 150,000 thousand years previous to us. It can be said they are our cousins though evolved much earlier. It is very logical therefore that if we can prove that Neanderthals had no oral or compositional speech, we can certainly say that *Homo sapiens* is the first species to have oral and compositional speech. Chapter 5 is all about that.

Homo sapiens or modern man migrated out of Africa, is the most accepted theory of migration of modern man. This is also called theory of monogenesis. There is another theory also prevalent amongst scientists, called multiregional migration.

According to this theory a species previous to *Homo heidelbegensis*, named *Homo erectus* had migrated out of Africa in groups, and wherever they migrated *Homo sapiens* evolved from them at many places and language also evolved at multiple places. Answers to all these questions will be found in Chapter 6.

Chapter 7 will discuss the concepts of many authors prior to present theory presented by the author of 'The origin of speech and language.' The smallest unit of speech is phoneme.

Some authors have not given consideration to phoneme but think of origin of word as a first step in the origin of speech. Their views are discussed in Chapter 8.

Chapter 9 discusses in detail the thought of Dr VS Ramachandran, about the origin of word and language. As stated previously he is the only neurologist or we can say only medical scientist who has placed his opinion in this field.

Chapter 10 is devoted to the 'Theory of evolution of speech and language' as it is presented by the present author.

New Investigative Techniques

1. We know radiography is X-ray photography. Cineradiography is a technique to record the movements of organs and systems within the body by having rapid exposures. The technique was evolved by Janker in Germany in 1930. Many scientists after that worked on it latter and the method came in clinical practice by 1953. The movements of the organs of articulation, specially of tongue during speech are invisible. They became visible though, by this technique. Figure 1 shows the movements of tongue.

2. *Electromagnetography (EMG):* While some neural activity is going on in the brain, a very low magnetic field is produced in that area, this can be tapped from outside. If there is excessive movement of the magnetic field in a particular area we can deduce that the particular activity has got its origin in that area.

3. *Electromyography:* Records the movements of the muscles in a particular task and with what strength they are acting. This indicates what muscles are acting in a particular activity and with what strength.

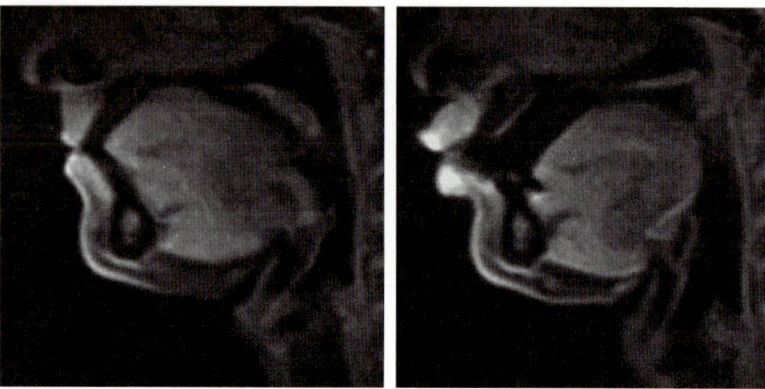

Fig. 1: Movements of tongue

4. *Functional magnetic resonance imaging (FMRI)*: Every one is aware of MRI. This is based on the principle that if a neural activity is taking place in a particular area of the brain the blood supply of that area is increased. Using an appropriate imaging sequence this can be traced and thus human cortical activity can be observed. By this technique which particular activity is taking place in which particular anatomical area of the brain can be traced. This way motor activity in Broca's area during speech can be traced.

5. *PET*: A PET scan uses radiation or nuclear medicine imaging, to produce three-dimensional color images of the functional processes within the human body. PET stands for positron emission tomography used to detect a health condition or functioning of the various centers in the brain. A radioactive material is produced and then it is tagged to a natural chemical such as glucose, water, or ammonia, which is known as radiotracer. The radiotracer is injected in the body with the nuclear medicine. If a particular activity is going on in the brain that area consumes more oxygen and the nuclear medicine concentrates there in the brain. This way, site of the functioning of a particular activity can be traced. The Fig. 2 will give an idea of hearing related activity in the brain.

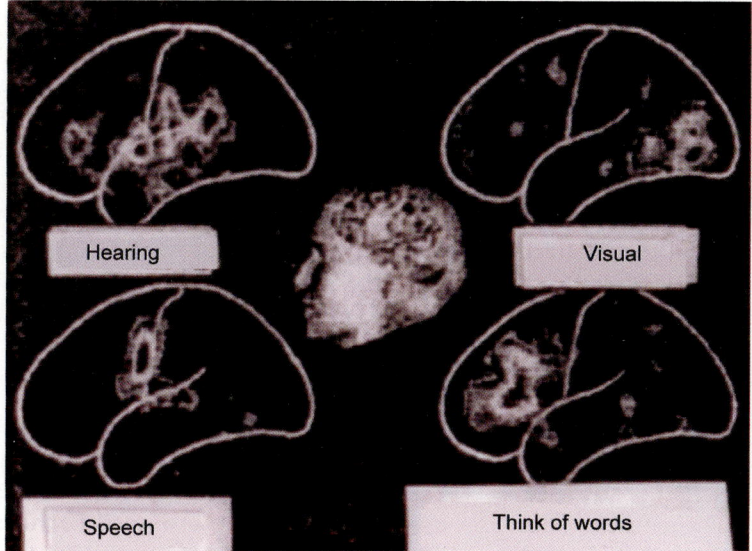

Fig. 2: Speech centers in brain PET

DATING METHODS

These are the methods to find out the age of the archeological material excavated during particular archeological sites.

Potassium-argon dating: Potassium-argon dating method or K-R is a radiometric dating method used in archeology. Potassium starts acuumilating in the molten rocks which starts solidifying from the molten lava of a volcano. Potassium 40 (valency) starts decaying into argon 40. The time since recrystallization into argon 40 is calculated by measuring the amount of argon 40 accumulated to the potassium 40 remaining. The long half-life of potassium allows the method to be used to calculate the absolute age of samples older than a few thousands of years. Half-life of the material is the time required for converting half of the material into other just as K to A, or decay of the material into another material. The dates of the archeological material are calculated by calculating the age of the rocks above and those below the sample, which can be a fossil or an article such as stone tools prepared by previous species.

Half-life of potassium 40 is about 1250 million years. The age of the rock can be calculated from, how many atoms of potassium 40 are decayed into argon 40. Due to the long half-life, the technique is applicable for dating mineral and rocks more than 100,000 thousand years old. It plays an important part in archeology. One archeological application that can be quoted is bracketing the archeological deposits at Oldowan river Gorge by dating lava flow above and below the deposits. It has also been indispensable in other east African sites with a history of volcanic activity such as Haldar, Ethiopia.

Radiocarbon Dating or Carbon Dating

Radiocarbon dating sometimes simply known as carbon dating is a radiometric dating method which uses the naturally occurring radioisotope carbon 14. Normal valance of carbon is 4. The carbon dating is used to estimate the age of organic remains of the archeological samples which are bearing radiocarbon. **They are reported as radiocarbon years before present (BP), present being defined as 1950.** When plants fix atmospheric CO_2 into organic materials such as starch during photosynthesis they incorporate a quantity of C14 that approximately matches the level of this material in the atmosphere. When the plants die or they are consumed by the animals or humans the carbon factor of this organic material has a fixed amount of radioactive material C14. Comparing the remaining C14 fraction of a sample to that expected from the atmospheric C14 allows the age of the sample to be estimated.

The technique of the radiocarbon dating was developed by Willard Libby and his colleagues at the university of Chicago in 1949. He has been awarded the 'Nobel Prize' in chemistry for this work.

Magnetic Reversal Dating

The mother earth has got a north and south geographic poles. Similarly, it has got north and south magnetic poles. The earth's north and south poles are fixed but the magnetic poles are changing. Today's north pole may change to south pole and *vice versa*. This happens after thousands or hundreds of thousands of years. How many years after and why it happens so is not still clear? But when it had happened before can be ascertained. The lava flowing during the volcanic eruption contains a few iron particles. When the lava solidifies the iron particles inside get fixed. They are fixed showing the north and south magnetic poles of the earth at that time. If when the earth's north and south magnetic poles were changed is known from that, the age of the volcanic rocks can be detected. If the age of the rocks below and above the rocks can be determined, the age of the fossils between the rocks can be decided. The age of the fossils at Atapuerca-TD6 site was decided as about 800 thousand years in this way only.

DN Lele

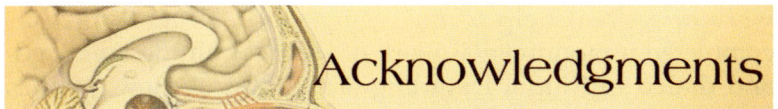

Acknowledgments

The present study is possible only due to encouragement I received from my teacher and famous otolaryngologist of JJ Group of Hospitals and Professor in Grant Medical College, Mumbai—Dr NK Apte. He himself was interested in writing a volume on comparative study and evolution of articulatory system of modern human. When he learnt from me that I am interested in study of evolution of speech in *Homo sapiens* he was delighted and encouraged me in this project and helped me in this present endeavour of mine. His project could not be completed as he passed away before completing it. His blessings will be always with this work.

Dr MN Gogate an otolaryngologist from Kolhapur, is very close friend of mine. His son Dr Nitin is in USA for a long time. He is interested in study of genetics. When he learnt through my friend about my interest, he was kind to send me a set of books in related disciplines. Perhaps that was the beginning of my present work. I am grateful to both of them, without Dr Nitin's help this work would not have seen the light of the day.

A renowned thinker, journalist and editor of many journals from our state Shri Kumar Ketkar is well-known in this country. He has interest in many diverse disciplines and has written columns in dailies. A few of them were about famous linguists Noam Chomsky and his disciple Steven Pinker. He is very outspoken and well-known critic in literature. I thought I will get correct advise from him. He not only appreciated my work but helped me by sending a few books on related subjects. These interactions and personal discussions on many occasions helped me in this undertaking. I owe a lot of respects to him.

I have special thanks, to offer to world known linguist Prof Ganesh Devy Not only he went through my manuscript but had valuable suggestions and supplied reference books which were of a great help to me in my work. Recently he has completed a monumental work of study of all the languages and dialects of India from linguistic point of view. He has undertaken an exemplary work in the field of tribal culture, languages and their lifestyle and other related studies, for which he has himself shifted his stay in tribal areas of Varora district of Gujarat. He is also of a great help to me in finding a renowned publisher for my work. I will be ever grateful to him.

I paid a visit to most respected linguist, Padmavibhushan Dr Ashok Kelkar. He was very glad to know that I am writing on the subject of 'The Evolution of Speech and Language'. He expressed to me that this subject being interdisciplinary has remained neglected by both, linguists on one side and medical men on the other. He gave me his best wishes and encouragement to complete the work. I will be always grateful to him.

My friend Dr Ratnakar Patwardhan, who is a voracious reader, specially in the field of physiology, functioning, evolution of brain and neurology in general, was of a great help to me. He has gone through the manuscript critically and suggested many corrections and improvements. Just expressing my sincere thanks to him will be quite inadequate.

My son Dr Pushkar, who is also an otolaryngologist and his learned wife Mugdha have always encouraged me. Mugdha helped me in translating some of the part from my Marathi version of the book. Dr Pushkar, who belongs to the same discipline and having special interest in "Phonosurgery" was of a great help to me. He has critically gone through the manuscript and had many suggestions. I am grateful to both of them.

I am working in the field of identification, education and research of hearing impaired for a long time. Here, my colleagues Pro Bele a renowned educational psychologist in the state. Principal; Aarati Khaladkar a renowned teacher of hearing impaired and President of Association of Teachers of Hearing Impaired, and our audiologist and speech pathologist Shrimati Krupa Menon, we all had discussions on many subjects related to the present treatise, these exchanges have helped me in clarifying many ideas and riddles related to the present subject I am very much grateful to all of them.

My elder brother Professor VG Lele who himself has written many books has always encouraged me and given lot of valuable tips. His blessings are always there with my work.

Our staff members Rupali Mahajan, Geeta Gavade, Dr Sunetra Rao, and Ananya were of a great help to me in my work. I am very much thankful to them. Special thanks to Rupali for her help in computer work.

Shri Shrikant and Deepali Nagare have helped in the art work of the most appropriate, meaningful and artistic cover design for this work. I am grateful to them for the same.

I will be also thankful to CBS Publishers & Distributors, special mention to Mr SK Jain, CMD and Mr Ramesh Krishnamachari of Mumbai Branch.

I request the readers to send their valuable suggestions at dattatraylele@gmail.com.

DN Lele

Contents

1 Historical Perspective

In the first chapter we will understand the difference between communication and language. Animals also, communicate with each other but by means other than language. There is no human community found which is without oral compositional language in the world and no other animal can communicate with speech. Many scholars of the world since ancient times have put forward their own theories of origin of speech and language. Greek scholars of ancient era Plato and Aristotle thought that the language is a God's gift to man, famous French scholar of 15th century Descartes and even 20th century linguist Chomsky were of the opinion that language is there since birth. In 18th century German scholar Max Muller who was a Sanskrit scholar also had put before some hypotheses and had facetiously christened them as hypotheses of origin of speech. All this fascinating history we will come across in this chapter.

The possession of speech is the grand distinctive character of man, and indeed it dwarfs the most other evolutionary achievements. The unique ability of communicating through an oral language clearly separates the human from animals. P. Liberman in his book 'Eves spoke' says communication via speech is uniquely human so much so that it often is used as the singular and most important dividing line between human and animal. Animals who talk are humans. In 1994, an article appeared in Time magazine titled 'How man began' within the article was the following statement 'No single essential difference separates human being from animals.' Yet in contradiction to such statement evolutionist offer that communication via speech is uniquely human. TH Huxlay in

1871 stated the same thing. Speech is the main dividing line between man and animal, all other differences are minor. It was thought that materialistic science is insufficient in explaining, not only how speech came about but also why there are so many languages.

The editors of 'The Cambridge encyclopedia of primate evolution' conceded that, there is no nonhuman language and no community deprived of language has been ever found.

Jean Aitcheson noted the "Evolution of speech and language" remains one of the most important hurdles in theory of evolution even in 21st century.

There is a famous Chinese poem composed 1000 years ago. It says 'you cannot visualize a mountain while you are on it, you can neither appreciate it from a distance, the mountain keeps changing with every new perspective. However, we are trying to solve the miracle of evolution of language, with the help of language itself.

Every living being tries to communicate with other members of his community. But there is vast difference between communication and speech. The honeybee communicates with its specific dance, which indicates its coworkers the site, distance and direction of source of honey. Male stickle fish changes the color of its belly to bright orange and indicates its mate, that it is ready for union, the frog indicates the same thing by its crowing and blowing out its neck sac. Chimpanzee and pre-human species communicated by vocal messages which cannot be called as speech.

Modern human and chimpanzee had a common ancestor. Since then many generations must have passed and many species must have evolved in between till modern human evolved. Our DNA sequence is 98% same as that of chimpanzee. Defining what it is to be human has for many researchers meant only to look at the 2% difference and assuming that in that small gap lies key to all the cities of the world, all the oil rigs, all of culture and language. We have got no living specimen of any species between our common ancestor and modern human. Of course that does not mean, that none of the species in between was not having any communication. They also must be communicating with each other with vocal messages. Gradually speech might have evolved in some species. We believe this miracle has not evolved from nowhere, we are going to find out how exactly this miracle happened.

We use language as much as we use air. The air has completely engulfed us. It is there around us right from the birth, same can be said about language. Above high mountains where air becomes thin and if by chance we are chocked we appreciate the existence of air, similarly we cannot imagine life without language. It will be a maddening situation. The infant after birth starts babbling in 2 to 3 months and utters first word at the age of one and half year. Since then it makes such a fast progress that we do not give attention to its origin. My father had speech, his father had speech, his father and uncles were talking.

This list may take us to our original ancestor who lived 200 thousand years back. We take it for granted that language is just there, as air is.

There is no single grand and convincing theory of emergence of language. Unearthing the earliest origins of words and sentences requires the combined knowledge of more than dozen different disciplines. It begins with a basic uncertainty about the validity of even studying how language evolved? The questions have taken different forms. Did language evolve at all? Is this question scientific? And even if it is scientific can we answer it? And whether it is worth trying to find out the answer.

Jean Aitcheson says the evolutionist also did not pay much attention to this subject. In 1866 in the founding statutes of the 'Linguistic society of Paris' perhaps the foremost academic linguistic institute of the time, a ban was enforced on subject of origin of language. The society did not accept papers either on the origin of language or the invention of universal language. This is amazing admission of defeat. 'The Royal linguistic society of London also had banned in 19th century to accept papers on the origin of language. Their opinion was, this is a waste of time. But you cannot sidetrack a question and solve it. Even the cognitive neuroscientists in 20th century had neglected the subject completely. In the 900 pages volume of Posner 'foundations of cognitive science' there are only two chapters on motor system, but not a single word about articulation of speech. Levelt feels origin of speech and language must be a stepchild of psycholinguists therefore they did not pay attention to it. A fourteen hundred pages volume of Gazaniga (2000). 'The new cognitive neurosciences' there is not even a mention of origin of speech and language.

Many or almost all scientists thought speech is an innate faculty gifted to man and is instinctive. Just as spider spins a web because it has got a brain (yes it has one) it can spin it by innate knowledge. We might be possessing speech in the same way. But one thing we forget that the spider spins a web but it does not know the Hooke's law of elasticity, it can only spin a web. Man knows about the Hook's law and can think of solving many other problems with that knowledge. For exchanging thoughts there must be a medium and that is language. But no one thought about how speech can be an innate property available to us. In his book 'Descent of man' Charles Darwin says even one accepts that an infant has got an innate capacity to speak no evolutionary will think that it came to man just like that. It must have been inundated in human brain by a little changes taking place step by step. But he did not mention about how these a little changes took place.

SOME OPINIONS ABOUT ORIGIN OF LANGUAGE

Jean Aitcheson and others are of the opinion that speech is an activity, which is achieved by human and no one has endorsed it to him. Small variations appearing step by step in pre-human species as an effect of environment and requirements to adjust with it, pre-human species made use of various motor activities already evolved in their time. New neural circuits were required to stimulate these activities and were evolved. As a result the size of the brain started gradually increasing. The species to maintain its primacy and prove its supremacy had to do various variations in their calls and perhaps speech came out of these efforts. But even then Jean Aitcheson could not explain exactly how speech originated out of these efforts.

Another evolutionist John McCrone put it this way. It all started with an ape that learnt to speak. Man's hominid ancestors were doing well even though world was griped in ice age. They had problems how to feed their rather oversized brain. Man happened on the trick of language and suddenly man became aware of this new vista and became self possessed.

Michael Corbalis stated in 2002. "My own view is that language developed much more gradually starting with gestures of apes and through the process of evolution what might have distinguished Homo species, the final switch was pushed on, and appeared a vocal language, from a mixture of gestural and vocal communication to autonomous vocal language embellished by gestures, but not dependent on it."

But he himself could not explain what was exactly the switch like, which introduced the transformation.

Evolutionist Carl Zimmer (2001) summed it up well, when he wrote, no one knows the exact chronology of this evolution because language leaves no traces on the human skeleton.

Jean Aitcheson admitted 'Holes still remain in our knowledge, in particular at what stage language leaped from some vocal calls to new element which human discovered being something which every newborn human is scheduled to acquire. *This is a puzzle. A 'puzzle' indeed*.

Under these circumstances many thinkers started to explain in terms of their own religious philosophy about origin of language.

MYTHOLOGY

Creationists strongly believed that language is a gift endowed by God on human race. God created world and gave birth to Adam on 6th day. He had created him in his own image. That is why he also started talking like god and understood what god conversed with him (Genesis 1:28–30). Within hours he gave birth to 'Eve' from the rib of Adam. This is perhaps the first mythological surgical feat in the world. Naturally she also achieved, the art of listening to God. God had created various animals before that. He asked them "Now you speak and call me your creator the God." But animals could not speak anything. Then God became angry and bestowed on them that 'You will be all governed by man and they will eat your meat.'

Vedas

Animals could not talk was the faith of creationists which was overcome by other religions. Vedas were created by the God of the gods Brahmadeva. The words of the script were filing the spaces for millions of years. He gifted the scripts to vedic rishis. That is why vedas are supposed to be written not by any human being. Vedic rishis only put the scripts in lyrical form. Some of the richas (Stanzas) are supposed to be composed by 'Sarma' the dog of gods and some are even composed by a bird 'Jarita and her offsprings (10/108 and 10/142).

Jain philosophy says 'Ardhamagadhi' the language of their scriptures is the oldest and even animals and birds could understand it. There are paintings in 'cave paintings of Ajanta' showing that even the animals and birds are also attending the discourses of Jain munis.

Buddhists say that Pali is a language of common man who can understand the scriptures of Bodhisattva. By them Pali is supposed to be, older than Sanskrit.

Medieval Period

Medieval historical period is also full of such stories. Chieftains of various states had curiosity to know which was the first language of the world? Greek historian Herodotus has told a story of Egyptian monarchs matic, 1st Pharaoh. He raised two boys under the care of a deaf and dumb caretaker. He wanted to know which language they end up with speaking first. When the children were brought to him after years one of them said something that sounded like 'becos' the 'Phrigian 'word for bread. He concluded, Phrigian was the first language. King James of Scotland Fredrik II and Mogul emperor samrat Akbar in 16th century are said to have undertaken similar experiments.

Renaissance

After the industrial revolution of 15th to 18th century in Europe there was explosion of scientific thought and culture. Many thinkers started exploring the possibility of understanding the secret of origin of language on the scientific basis. Some were exploring the role of brain in many physiological activities and also in the production of language. World famous linguist Noam Chomsky put forward a theory. He says the fact that human child acquires such a vast knowledge of language in early years of life indicates that it must be there prior to birth. There is probably a 'language organ' present in the brain and it is innate. He was influenced by the great Greek thinkers like Aristotle and Plato. Plato claimed that the very thing, that human has got knowledge of the world in such depths, is not possible to acquire by external stimuli. This must be there prior to birth as an innate gift. Chomsky claimed that human child has got understanding about sounds of speech and language and syntax from the birth, he called it 'generative linguistics' and all the complicated syntactic arrangements as 'generative grammar'. But the question was raised, how can the child have the knowledge of so many languages present. He replied at the time of birth the child has got understanding of universal grammar, it later accepts only that which is related to the language spoken around it. We will see more about Chomsky's theory about origin of speech latter in Chapter 7.

Thus, far we have seen different opinions about this miracle, of origin of speech from the point of view of different thinkers. Different thinkers were looking at it as different sounds used in language or phonology and the words formed from the combination of these sounds. How human came around to decide what word to be used for different objects or actions taking place around him? Plato had put forward a solution to this problem in his book 'Cratilus'. There he says that everything has got a right name of its own, which comes by nature, and that name is not whatever people call it by agreement but there is a kind of inherent correctness in the name. Socrates eventually reached a conclusion that the names are to represent the things, it must be in virtue of the fact that they possess some internal likeness to the things that they are to represent.

Premodern Period

Herder (1744–1803) who might be credited as a founder of the science of origin of language. He was pupil of Kant. His essay on language published in 1772 was an attempt to refute the hypothesis of divine origin of language by proving that man developed language in tandem with his development of reason. In the middle of 19th century a great contributions were made by Max Muller (1823–1900) and William Whitney. Muller facetiously christened many hypotheses of origin of speech. Let us see some of them.

Max Muller's Theories

1. *Ding-dong hypothesis:* There are many metallic and similar objects such as a bell, a rolling stones, etc. There are also sound making events in the nature such as roaring of clouds, sudden splash of lightening. There is some natural and inherent connection in these sounds and the events and objects. Therefore similar words were used for them.

2. *The bow-wow hypothesis:* This is based on onomatopoeia principal and that is the foundation of speech. Words are composed of sounds which imitate the objects. The sound accompanying the blowing of the wind, incessant rain, birds songs are imitated and matching names are given to these events or objects. But what about the objects which do not make any sound, or the event of the past, abstract things or concepts. These were some of the objections raised against this hypothesis.

After him many linguists have proposed many theories thinking on similar lines. These theories were grossly incomplete and could not stand to scientific reasoning. This was perhaps the reason the linguistic society of Paris and Royal linguistic society had put a ban on papers presented about the origin of language. Such a ban did not deter the scientist from thinking to satisfy their curiosity. Famous Indian linguist Prof. Gajendragadkar has said "someday someone will come and say, "well I have solved this puzzle."" He had perhaps thought of the story of Dharmaraja and Yaksh from famous epic Mahabharat about 10 questions of Yaksh in mind.

There might be some truth in each theory but that may indicate how the language was enriched in different ways and not how to solve the puzzle of its origin.

Conclusion

Nobel laureate Tinbergen defined the speech as 'speech consists of rapid stream of sounds, extended in time, that we produce with the vocal apparatus' But how are these rapid sound sequences produced? And what orders them? Were the two questions still unanswered? Perhaps it might be so that most of the articulatory movements are invisible and functions of brain were only to be judged by the effects of the lesion or after postmortem studies. Since the invention of new techniques of investigations such as PET, FMRI, etc. we have learnt more how exactly the articulatory system works, the functioning of live brain, and since then many scientists have taken interest in the subject of origin of language. P. Liberman, Steven Methane, Dr VS Ramchandran are some of them. After going through these different studies whether one can propose a coherent theory, based on logical thinking, after filling in the gaps and removing the drawbacks, is the subject matter of this work of mine.

We will have to answer also, since when the language is there, during this process of evolution of modern human, from the first pre-human species Australopithecus. Each species had improvement in anatomical and physiological functioning which at some time helped in the production of this miracle of speech. To find out this chronology we will have to study the process of evolution from Australopithecus, the first pre-human species to modern human or *Homo sapiens*. Let us take a survey of it in brief in the next chapter.

Evolution of *Homo sapiens*

Second chapter starts with Charles Darwin's theory of evolution by natural selection. This theory is now well accepted by scientific world. From unicellular species like amoeba to modern human are all evolved by variation, and its acceptance. A few changes here, a few changes there, new capacities were acquired and a new species evolved. In the same evolutionary process some day the ancestor of chimpanzee and human was evolved. Passing through many pre-human species modern human evolved. All the pre-human species were communicating with each other by vocal messages. They had no speech. Why? We will have a brief review of this evolutionary process up to modern human and try to find the answer in this second chapter.

CHARLES DARWIN AND HIS THEORY OF NATURAL SELECTION

"The origin of species by means of 'Natural selection' or the preservation of races in the struggle for life."

Publishing a monogram of this unusually lengthy title in 1859, Charles Darwin shook the scientific world.

Charles Darwin was born in 1809. His father was a medical doctor, grandfather was also a medical practitioner, naturally his father thought of educating his son in the same field. After completing his schooling he admitted Charles in medical college at Edinburgh. Charles really never wanted to be a doctor of medicine. He quit the college and came back. His hobby since childhood was collecting butterflies, collecting variety of insects, and keeping notes after observing them. His father then thought of sending him to a church, for training of priesthood. He could not settle there too. At about the same time, a naval captain Mr. Fitzirald was going on a naval expedition. His intention was to see new world, exchange goods, and earn some money. He wanted a friend to join him on his trip. Charles offered himself for this trip, to go with him as a companion. They started on their voyage in a small vessel named 'Beagle' in 1831. There ship Beagle went round the Southern tip of South America and reached the islands of Galapagos off the west coast of South America in 1834. These islands are formed by volcanic eruption and are about 1000 km from the coast of Ecuador. The flora and fauna of the islands was entirely drifted and reached the islands from the mainland and therefore was the same on all the islands. Charles, however, found that beaks of Finch birds on different islands were different. Why it was so? If they had all came from the same destination and were belonging to same species the difference needs explanation. He gave a deep thought to this observation. He found that the fruits of tribulus, which was the main food of the finch birds on different islands' were different than mainland. On the island where the fruits of tribulus were soft and brittle, the beaks of birds were long and sharp on the other hand, on the other island where the fruits of tribulus were hard to crack the beaks of the birds were short, deep and the musculature attached to it was also strong. It was therefore easy for them to crack these hard fruits. Thus, variation took place in the characters of the birds on two islands. As this variation was beneficial for their survival by getting their food easily it was accepted by them and carried in the next generation. In few generations to come the change became genetic, and a new species of Finches was evolved. Thus, he observed that the 'natural selection' of the variation in individuals, which was beneficial, is the cause of appearance of new species. Thus, two different species of finch birds were evolved. Here the roots of his famous theory of evolution by natural selection were established. He further continued his observations on different animals and plant species for further 25 years, and after conforming his line of thought published his world famous volume in 1859, "The Origin of Species By Means of Natural Selection". Thus, he thought, from unicellular amoeba to modern human all the living world must have evolved over 4 billion of years, through the same process of 'natural selection.'

Darwin had four principles of 'Theory of Evolution by natural selection.'

1. There is always a variation in each individual of a species, which is known as 'A variable.'
2. If this variable is beneficial for adaptation to environment, and for survival and reproduction, it is carried further in next generation.
3. In each species more number of individuals are born than, number which will survive. All of them have to go through a struggle for life. The mate always selects the male as a partner, who has got better chances of survival and reproduction for continuation of species.
4. In some thousands of generations the changes are adopted by future generations.

They show changes in external appearance and functioning of the systems, thus a new species is evolved.

Sometimes variables can appear because of other reasons. The environmental changes, paucity of food or increased population of predators, are some of them. Under these circumstances same principle applies. Those who have got survival merits only survive. Mate chooses such males only as partner and a new species having better survival chances is evolved.

Credit of exploring fossils for the first time goes to French scientist George Cuvier (1769 to 1832). He was the father of the science of paleontology and comparative anatomy. These sciences help us to know about lifestyle of the animals which lived on the earth one time but are now extinct.

Bupho, Buckland, Cuvier and many others tried to analyze events in nature. No one still dared to contradict the religious scriptures. They tried to connect all the evidence with the scriptures.

Believers of course staunchly refuted theory of evolution. They still believed that god has created this universe. Still even today, there are pockets of communities having faith in this belief. Those accepting and propagating Darwin's theory had to undergo many hardships. Sometimes even had to sacrifice their lives. In every religion there are somewhat similar myths about origin of life on the earth. Even Darwin had to tell the truth of his theory disguised under very intelligent explanation. He said "I have not stated in this book how man was evolved. I have not criticized Genesis or Bible. I have only noted my observations and the inferences. Evolution is a subject related to very ancient period and therefore there might be such impressions drawn from my book." Even then as the subject matter of the book was going against the Bible he had to face very severe criticism. In spite of this, many editions of his book were sold and are still appearing.

'Natural selection does not take place with an intention of evolving a particular new species.'

How the living being might have evolved from the inorganic world? Darwin had put down his own thoughts "By chance combination, as a result of effect of cosmic rays or some such thing, from the inorganic matter in atomic form, a protein molecule was formed which was not only capable of surviving itself but also producing identical molecules again and again. By some distinctive combination of these molecule the first unicellular species evolved." In Darwin's time the DNA molecule was not identified but the molecule he thought, giving birth to several molecules, identical in configuration can be designated as DNA. This event took place perhaps about 4 billion years ago. The life must have begun under water. From unicellular life, more complex organisms must have evolved by 'natural selection.' Then appeared vertebrates, fishes which could survive under water and next species which could live underwater as well as on land and therefore called, amphibians. Latter fully terrestrial animals developed which were reptiles. From the reptiles, on one side evolved the birds, and on the other the mammalians. The sequence of evolution is not imaginary but there are proofs available.

In 1861, fossils of a specie, were found in Germany, which is classified as 'Archioplatis.' Some of its appendices are showing similarity to reptiles and some to birds. In Australia, there are birds having flat beaks like ducks, and one more species 'spiny ant eaters' both of which lay eggs like birds and the mother feeds the offsprings like mammalian mother. The Nobel award winner Francis Creek once said "By his theory of natural selection Darwin has given us the secret of life." Stephen J Gould says evolution does not take a course of single line. Tree of life bustles with stem, boughs (branches from the bottom of the tree) and branches leading to different species, or dead end.

EVOLUTION OF *Homo sapiens*

The last to evolve was our species of *Homo sapiens* or modern man to which we belong. We are still able to find many of the species on the intermediate steps of evolution because those have found an ecological nitch comfortable for their survival and reproduction. This is the reason, why we can still have intermediate species like frogs, reptiles, birds and domestic, as well as wild animals. They are helping us to study the Darwin's theory of evolution by natural selection.

Darwin put forward one more thought. Nature is very clever and miser too. If by the use of same organs which are already in existence some other function can be undertaken, or exapted (borrowing), nature takes that opportunity. When the man developed the art of communication by oral speech, instead of evolving some new appendices for that function, nature put in use the organs which were functioning for eating, drinking, sucking in short for swallowing food in different forms. They were brought in use for this new art of speech. Their old function still continued, but the coordination of the muscle group was changed for the new function. Naturally new neural circuits were developed in the brain for this new function.

The respiratory system was originally evolved for taking in fresh air rich in oxygen and expelling out air polluted with carbon dioxide. It was now brought in use for supplying energy in the form of puffs of air for the production of speech. Organs were the same, their original function was retained, but they were now put in use to perform the new function of speech production. This is called 'Neo-darwinism' or 'descent with modification' existing organs were brought in use for some other function. Gould and Vergha (1982) have coined the new term 'exaptation' to describe the process in which there is borrowing of the activity of same old organs for new function. Sometimes two or more attributes existing are adapted for performing a function which is something more than their simple sum.

Was our existence unavoidable or inevitable? Evolutionary biologist do not see that evolution of any of the species is unavoidable or inevitable and intentional. Rather they see it as a long process of events each depending on the other, and each unpredictable and unique. Most human features did not evolve in one step nor together or as a one package. What is more is, the functions of various features are changed to entirely some new function as sited above, fingers previously used for holding and climbing, branches of a tree, now played piano.

Primates

Scottish zoologist Linneus did a great work of classification of animals according to their external appearance, anatomy and functioning of system of organs or physiology, lifestyle, food habits, etc. This is known as 'taxonomy' Linnaeus classified apes and pre-human races in one big class known as primates. Primates in Latin means 'chieftain', or chief of all. While assigning this name to this class, Linneus had in mind the modern man which belongs to this class and is superior to all. Ancestor of chimpanzee and human is the same. From this common ancestor on one side evolved the chimpanzee and successive species of big apes such as gibbon, gorilla, etc. On the other side evolved the pre-human species and modern human in the end. First to appear in this class was designated as astralopithecus, astrolo means belonging to south and pithecus stands for, 'like ape.' This event took place about 6 to 7 million years ago. No fossils of our ancestors are ever found. The archeologist have come to the conclusion that the ancestor of chimpanzee and human was the same by studying the genetic configuration of chimpanzee and man. There are many similarities in chimpanzee and human on molecular level. Our blood groups are the same. Many of the proteins, hormones,and enzymes organized by DNA are also similar in molecular structure. Species Chimpanzee is still existing and available for study, from which some of these conclusions are drawn.

Unfortunately no specimen of any of the pre-human species is alive today. We can study their appearance, anatomy, life-style, etc. from their fossil remains and the stone tools left by them around, which are finds of excavations undertaken by archeologists. This way we can know whether they were quadrupeds, dwelling in trees or were bipeds like us. What were their food habits? How they were communicating with each other and so on? This is known in technical terms as 'anthropology.' We have already seen how dating of fossils and articles found around them is done.

Out of pre-human species, the oldest species explored is astralopithecus chadensis. It was evolved probably about 7 million years ago. After that Astralopithecus Egyptopithecus evolved about 5 million years ago. Only a small number of fossils of this species are available, therefore very little can be said about them. Archeologist Mary Leaky discovered a fossil near Turkana lake in Kenya in 1995. It was thought to be 4.1 million year old. She named it astralopithecus Africansis, Tim Wright from Berkley university of California has excavated a fossil which is supposed to be oldest in the astralopithecus group, and is in a better condition. It is supposed to be 4.4 million years old. In Ethiopia, there is a community known as 'afar.' The fossil was found near their habitation and hence it was designated as astralopithecus afarensis. One of the fossil in that area is found in a reasonably good condition. The excavator of this fossil Donald Johanson named it 'lucy.' It is estimated 3.6 million years old.

Astralopithecus was perhaps able to walk a little on legs like apes. His legs were bent in knees outside, therefore it could not walk fast or run. They were arboreal (tree dweller), they built a nest on trees. Height was about 1 to 1.5 meters. Configuration of toes indicate that climbing or hanging from trees was perhaps easier for them. Their forearms were longer than legs. Their face and jaws were protruding out, ahead of forehead. They Had big size teeth which indicates that their food was mainly vegetarian. Dimorphism was clearly present as the male was larger than female. The size of the brain was about 400 to 530 cc. **As the brain matter is obviously decayed and lost over the time, it is customary to give the size of the brain equal to the capacity of the skull cavity in cc.** All primate brains have a distinct sylvian fissure separating frontal and parietal lobes from temporal lobes of brain. Astralopithecus are included in genus hominid.

Figures of skulls of all pre-human species and their information is given in Table 2.1.

Homo habilis

Fossils from the period between 2.5 and 1.8 million years are found in the gorges of the river oldovan. Their brain size was about 674 cc. Brain size of the fossils found in the delta of kubi river or delta kubifora was about 775 cc. They were structured somewhat like a human being and partly like an ape. They were classified in the genus 'Homo' and were designated as *Homo habilis*. Their teeth were smaller in size indicating that there was meat in their diet. As their brain size was bigger, it needed more energy and therefore non-vegetarian diet was essential for extra energy. For the first time stone cut tools were found in the period of *Homo habilis*. The appearance of tools indicates that their cognitive knowledge in one domain at least was evolved. Naturally more neural circuits were formed in the brain. Whether *Homo habilis* were capable of hunting themselves is doubtful. Even though they were partially biped they had no capacity to run after a prey. Astralopithecus and *Homo habilis* were partially arboreal. They were perhaps scavengers consuming carcasses of animals killed by larger predators. To bring the carcasses from a distance, to cut them in pieces, and distribute, it was necessary for them to live in a group. For having a comfortable nice life in a group it is necessary to know what is going on in each others mind. This is known as 'theory of mind' and is the function of 'Mirror-neurons' in the brain. We will discuss more about these mirror neurons later, as they play very important role in origin of speech.

Table 2.1: Figures of skulls of all pre-human species

	Astrolopethicus afarensis	Homo habilis	Homo erectus	Homo heidelbergensis	Homo neanderthalesis	Homo sapiens
Height (mt) physique	1–1.4 Light build arms longer than legs curved fingers, toes	1.5 Robust but human skeleton	1.3–1.5 Robust but human skeleton	1.5 Robust but human skeleton	1.5–1.7 As before but adopted for cold	1.6–1.8 Modern skeleton but adopted for warmth
Brain size (ml)	400 to 500	600 to 800	750 to 1250	1100 to 1400	1200 to 1750	1200 to 1700
Skull form	Low flat forehead, projecting face, thick grow ridge.	Larger flatter face	Flat thick skull with large occipital and brow ridge	Higher skull: Face less protruding. Similar to *H. erectus*; teeth may be smaller	Reduced brow ridge: Thinner skull—large nose midface projection. Similar to *H. heidelbergensis*; teeth may be smaller.	Small or no brow ridges high skull.
Jaws teeth	Large incisors and canines, moderate sized molars.	Robust jaw; large narrow molars	Robust jaw; smaller teeth than *H. habilis*			Shorter jaw; chin developed teeth smaller
Distribution	Eastern Africa	Eastern Africa	Africa; Asia and Indonesia	Africa; Asia and Europe	Europe and Western Asia.	Africa and Western Asia
Known date million years ago	4 to 2.5 millions of years	2.4 to 1.6 millions	1.8 to .3 millions	0.4 to 1 million	0.3 to 0.15 million	0.2 to 0.15 million

We are included in the class of vertebrates. We have got a column of vertebrae on the back supporting the body. This is called vertebral column (Fig. 2.1). At the center of the vertebral column there is a canal starting from lower end and entering the skull high up. In this canal there is a rope-like structure called spinal cord. Spinal cord is made up of nerves coming from and going to various organs and muscle groups of the body. The vertebral canal is attached to the skull and at that position there is big hole in the skull cap called 'foramen magnum.' Spinal cord enters the skull through the foramen magnum. In quadruped (Fig. 2.2), the foramen magnum is on

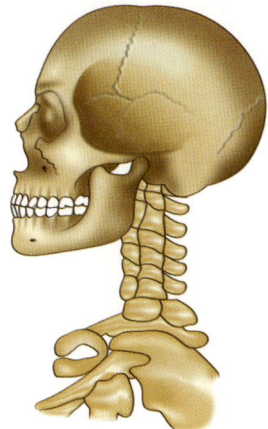

Fig. 2:1: Biped vertebral column going in from below

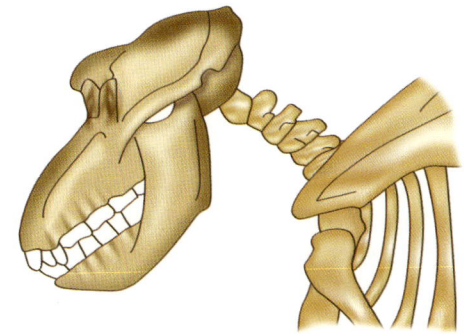

Fig. 2:2: Quadruped vertebral column going in from behind

the back side of the skull and spinal cord enters the skull from that direction. When Homo species started becoming biped the foramen magnum started shifting forward, as the vertebral column shifted forward and the spinal cord within also enters the skull from below instead of from behind.

What changes this forward movement of vertebral canal brought about in the configuration of the anatomy of mouth, food pipe and respiratory system? We will see in detail those changes in the next chapter. Some of the parts of the brain, when they increase in size by addition of new neural circuits, make an impression from inside on the skull cap. The position where there was going to be future Broca's area such an impression was seen in case of skull cap of *Homo habilis* indicating that there was the beginning of the formation of Broca's area. We will discuss the function of Broca's area in relation to speech in next chapter. As the teeth were smaller the jaw was also receding backwards, though it was still forward than the forehead.

Did astralopithecus and *Homo habilis* had vocal communication? Chimpanzee had vocal communication. They were communicating with each other by there harsh voice which was called as grunts, and barks. The pre-human species, are progressive steps in evolutionary process therefore they also must be having a vocal communication. How this vocal communication progressed and changed, will be interesting to learn as human language is the final result.

Homo erectus

Some more fossils were found in 1984 which belonged to a period between 1.8 and 3 million years old. A particular fossil found of a boy of sixteen years old at Nariokotome in Kenya was relatively in a good shape. That must be about 1.6 million year old. He must be having a capacity to walk on two feet fast. The walk of Australopithecus and *Homo habilis* was on curved toes or knuckle walk. Toes of this new species were different. They were straight and the big toe was, in line with other toes. While walking they use to touch the ground with heel first or what is called 'heel strike.' The pelvic and thigh bones were more or less like modern human turned inside, and those individual must be able to stand erect with legs brought together. This species was therefore designated as *Homo erectus*.

In inner part of our ear there is an organ called labyrinth (Fig. 2.3).

Its function is to maintain the balance of the body. It consists of a hollow sack called vestibule to which are attached three semicircular canals. The canals are placed in three different directions at right angle to each other. The hollow space and the canals contain a fluid. With any movement of the body, the fluid in the canals also moves, with the movement, the nerve endings floating in the end of the canal are stimulated. This stimulus is carried to the brain. The brain immediately orders the musculature concerned to act and correct the balance.

Spur has undertaken the CT scan examinations of the skulls explored and demonstrated that the direction of the canals in *Homo erectus* was like that in modern human as they had become almost biped. The length of the arms of *Homo erectus* was shorter than the legs. They had become biped and while walking arms were free to pluck fruits, to bring back the carcass of the hunted prey to the cave for their little ones and their mothers.

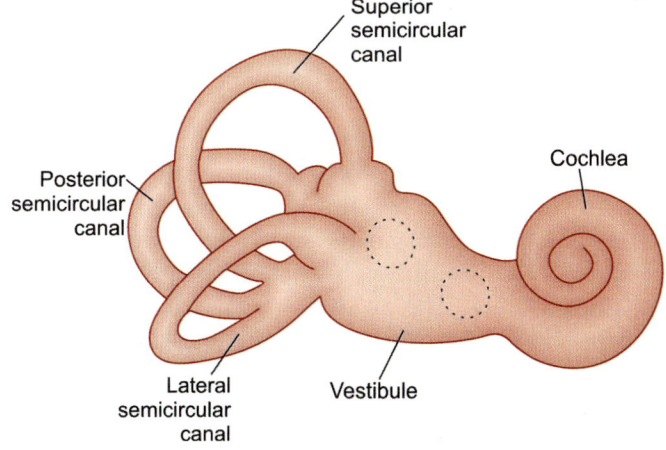

Fig. 2.3: Labyrinth

As their head was now away from the ground because of the biped position it was exposed to less heat than the rest of the body. High up the wind also blows faster which reduces the temperature of the head. Just by covering the head by leaves he could protect his head from heat. It is proved that even if the temperature is lowered by two degrees the efficiency of functioning of the brain can improve.

Because of bipedalism and free hands he could carve better and sharp weapons. Sharp ended stones or stones with sharp edges like an axe, to cut pieces of meat, tendons, or branches of trees and to make sharp ended weapons like spears.

As a result of bipedalism, the vertebral column was shifted forward so also the foramen magnum. Teeth were small and therefore the jaw size was also small and it was receding backwards. The larynx was pushed from behind by vertebral column and from in front by the jaw, as the larynx had no other direction than downwards to shift, it shifted downwards in the neck. Along with it, the tongue was also pulled backwards and downwards. The back of the tongue or the posterior face of the part of the tongue pulled downwards was now facing backwards. A new organ cavity was formed behind the tongue called pharynx. The pharyngeal cavity extended upwards to the base of the skull. The part behind the nose came to be called nasopharynx and the part, behind the larynx below, as laryngopharynx. The laryngopharynx below opened in the food pipe or esophagus. The respiratory system, which starts from the larynx was in front of esophagus.

Larynx performs one more function in monkeys and apes and was performing in pre-human species which were not bipedal such as astralopithecus and *Homo habilis*. When they had to climb a tree holding a branch or hanging from it, they needed more power. The accessory muscles attached on one side to arm and on the other side of the chest wall. Chest wall was fixed by deep inspiration and closing the glottal opening tightly, not allowing any air to go out. This way the muscles could use all the force only on one side to climb. During the time when *Homo erectus* were evolved they were biped, and climbing a tree or hanging was not necessary for survival. There was no necessity of closing the glottal opening tightly. For closing the glottal opening tightly. The vocal cords had to be of a firm material like cartilage which was no more necessary and the cords gradually became membranous. This made an important change in their vocal calls. The vocal calls in Astrolopethicus and *Homo habilis* were harsh and rough. While the quality of the calls in erectus onwards became more melodious. Bipedal walk had already given a rhythm to all the physiological activities including the vocal calls. Vocal calls therefore, with rhythm and melody acquired special musical quality.

Homo erectus became bipedal and as a result their pelvic bones and thigh bones turned inside. The result was the birth canal became narrow. It was difficult for the full size fetus to come out. Fetus therefore in erectus and all further species was delivered premature. The head of the fetus was smaller in size and the fetus was not independent. This is the case in modern human too. The fetus needed help and care of mother till the age of 1 to 1½ year. The brain size of the fetus is about ¼ the size of an adult. It grows fast and acquires the adult size proportionate to the body size by the age of 7 to 8 years. Till the infant is dependent the male had an additional duty to take care of mother and infant. Because of bipedalism his hands were free. He could go long distances for hunting and could bring the carcasses back to the cave. For all these additional activities and increased repertoire of the vocal calls, more neural circuits had to be developed in the brain, and therefore, brain size increased to about 850 to 1200 cc. For all these new activities erectus had to live in a community group.

Glyn Issac and his group have found such a community base in the delta of kubi river in East Africa. This habitation was named as FxJj50. There they found about 1500 pieces of stones of different sizes for preparing stone tools. Bones of about 20 different species of animals such as giraffe, zebra, deer, monkeys, etc. were also found around the habitation. Obviously they were hunting on a very large scale. There is no evidence that they were using fire, there is no constructed fire place found at their camp base. They were living in caves to protect themselves from cold and rain. They were preparing stone tools but with the same skill they never thought of preparing any symbolic articles. Their vocal call repertoire might have been increased in number but still they had no language. *Homo erectus* were living in community base. When the population increased they started migrating out of Africa and spread all over the globe. Their fossils are found as far as, Narmada valley in India, in Pakistan and China. Fossils of *Homo erectus* are found on an island in Indonesian group of islands. *Homo erectus* were there till 12000 years ago and then they became extinct.

Heidelbergensis

Some more fossils were excavated at Box Grove in England. The detail study of these fossils indicated that they belong to a species which is advanced than the *Homo erectus*. They were named as *Homo heidelbergensis*. Similar fossils were also found in north Africa. Their brain size was about 1100 to 1500 cc. Teeth were of smaller size as meat was their main diet, the jaw was also small but still it was protruding in front of the forehead. Their tools were much advanced. They were more like *Homo sapiens*.

From Heidelbergensis a new species evolved in Europe. It was designated as *Homo neanderthalesis* as the original fossils were found in Neander valley in Germany. This species evolved about 250 to 300 thousand years ago. From Heidelbergensis in Africa about 150 to 70 thousand years ago a new species evolved which was designated as *Homo sapiens*.

Neanderthals

Neanderthals were existing on the earth up to thirty thousand years ago. Further no other species evolved from neanderthals perhaps it was evolutionary dead end. The tool making method of *Homo erectus* and *Homo heidelbergensis* was different, it was known as oldovan industry. While the method of neanderthals was a little advanced and was known as Livolice method. Brain size of neanderthal was about 1200 to 1750 cc. It was larger than *Homo sapiens*. Does it mean that they were cleverer than

Homo sapiens? No. Intelligence does not depend upon brain size only. It depends upon 'Encephalization Quotient' (EQ).

The average body mass of any species and their brain mass is taken into consideration in calculating EQ. The answer is derived by complex mathematical and statistical methods. The EQ of erectus and heidelbergensis is between 3.5–3.8. In neanderthals it is 4.8 and in *Homo sapiens* it is 5.3. Why then the size of the brain of Neanderthals is larger than that of *Homo sapiens*? In Europe, there was ice age. The brain size of animals living in ice age is always larger. The body size of the Neanderthals was massive as the muscles of the body were large and massive. To operate this larger muscle system more neural circuits and more energy is required naturally therefore size of the brain was also larger. The chest was barrel-shaped and the muscles of chest were also massive. The teeth were little larger than modern human and there was big gap behind the second molar which made it easy for the third molar to erupt. They had no chin. They were fair skinned.

Nerves carrying sensations and supplying the muscles of various parts of the mouth such as tongue, teeth palate, etc. come out of the skull through the various openings in the base of the skull. The 12th or the hypoglossal nerve supplying the muscles of the tongue also comes out of an opening. This opening of the 12th nerve is larger in *Homo neanderthalesis* and *Homo sapiens* than the previous species. As the tongue in these two species can do more movements and have more nerve supply, the 12th nerve therefore became thicker, and a larger opening was required for it to come out. In earlier species the opening is like that in apes. These conclusions were drawn by Richard Kay and his colleagues by measuring the opening in variety of specimen.

The second measurable canal was that between the two vertebrae. Nerves supplying thoracic muscles come out of these canals. The thoracic muscles in these two last species had to have more fine control for speech. Though the Neanderthal cannot have so much fine control because of its barrel-shaped chest, however, the *Homo sapiens* needed fine control as they had developed speech. Maggi-Cecchi and Collard found a stapes and other middle ear bones in a Hominid species. After undertaking CT scan it was found that the bones were of the same size as that in modern man and they probably had same range of hearing. In fact, we are using the same auditory system developed a million years ago.

Neanderthals were living in small communities. Their family life was restricted to one mate till the upbringing of one child. Next time the female may chose another mate. How much was the cognitive development of Neanderthals? Neanderthals had intelligence divided in three domains. About this we will be seeing in more details in Chapter 5. The ancestors of Neanderthals and *Homo sapiens* were the same. In a way we are cousins. Why then Neanderthals became completely extinct 30,000 years ago? One reason is the ice age was completely disappearing from Europe, and they could not stand for the changing environment. Another reason might be *Homo sapiens* had developed language. They could exchange thoughts and plan better strategies. Neanderthals could not stand to their attacks and survive. Even then the puzzle of complete extinction of Neanderthals is not still solved.

Homo sapiens evolved in Africa from Heidelbergensis, who had evolved in Africa. Very few fossils from the period between 150 to 250 thousand years back, are excavated therefore very little can be said from that period. About 100 thousand years ago at Skhul and Qafzeh in Israel the first fossils of *Homo sapiens* out of Africa are excavated. Auditory system of Neanderthals was almost the same as that of *Homo sapiens*. Only difference is, we have the capacity of decoding the phonemes of speech, the art that we have newly developed.

Most of the anatomical details of Neanderthals were almost similar as that of *Homo sapiens*. Alison Wray says, they had the capacity for segmenting the vocal calls. Does this suggests that this was the beginning of the language? If not so is there any evidence that they had no language and only *Homo sapiens* had it? We will be discussing all this in Chapter 5. Let us first find out, how the modern man speaks then it will be easier to find out how he developed this art?

How Modern Man Speaks?

HISTORICAL PERSPECTIVE

Plato and Aristotle in ancient era, before Christ thought that speech is related to mind and both were present from the birth. Descartes in 15th century and Chomsky the renowned linguist of 20th century were also of the same opinion. It was an innate skill. Before trying to find out the truth of these statements, we will have to know the physiological basis of the speech and language of modern human. After the industrial and scientific revolution or renaissance of 15th to 18th century the scientific thought progressed very fast. Many scientists tried to solve this puzzle of acquisition of speech by modern human. We will take a brief resume of this and will know more about, how exactly man can utter different phonetary sounds and words leading to language then only we can talk about the evolution of this physiological activity.

Once during the famous debate about Charles Darwin's theory of natural selection Benjamin Disraeli asked "after all who is man? Chimpanzee only, with some abilities acquired through natural selection as per theory of evolution of Darwin or an angel sent by God with some special abilities." "If we use today's computer language, is man a chimpanzee only, with some special software acquired by natural selection or a God sent angel with the components for speech and language which were never present in any primate or pre-human species."

Evolutionary biologist of course were of the opinion that we are cousins of chimpanzee, with further evolved abilities by natural selection and labeled as pre-human and human species. Anthropologist and evolutionary biologist believe chimpanzee and man have got the same ancestor who lived about 6 to 7 million years BP. On one side chimpanzee and further species of apes such as gorilla and baboons evolved and on the other astralopithecus and further pre-human species and lastly *Homo sapiens* or modern human.

Western Philosophers

Believers did not approve of Darwin's theory of natural selection. They still adhered to old scriptures and believed that universe was the creation of God. Somewhat similar was the belief of all those following some ancient religions of the world. The great Greek philosophers Socrates, Plato, Aristotle, all believed that man is a special creation of God and is endowed with special abilities that are innate. The 16th century French philosopher Rene Descartes was of somewhat similar opinion about mind and language. As a matter of fact Descartes is considered as the originator of the modern scientific thought. He compelled himself to doubt everything that can possibly be doubted. The end point of this thinking, was his famous conclusion " *I think*, therefore I am". (Possibly, he did not doubt the thought!) This thought of Descartes is known in scientific world as 'Cartesian thought' or 'Cartesian doubt.' He accepted Plato's thought about mind, that mind is existing from the beginning of life. It does not change or progress. Abilities of mind are there prior to birth. As language is the function of mind it is also innate (present from the birth). Descartes agreed with the thought that language was innate in man and because of that he is powerful amongst all. Speech is an external expression of language and is expressed by mind's order. Some species of parrot can also speak but they only imitate. They do not know what they are talking and cannot ask questions.

According to Descartes and later by Noam Chomsky man has a 'language organ'.

Richard Owen, founding father of comparative anatomy pointed out that the great gap between man and apes is due to an absent anatomical structure 'hippocampus minor' which was not only responsible for the advanced mental abilities of man but also the possession of language by *Homo sapiens*.

Evolutionary biologist did not approve of this theory. The presence of 'hippocampus minor' could not be confirmed by dissections and therefore not accepted by the great anatomists of his time, Thomas Huxley, and also by the modern biologist. Owen was wrong about hippocampus minor. However, Owen correctly observed that language is not a divine gift but is related anatomically with a center in the brain. This was compatible with Darwin's theory of evolution which stated, that our brain evolved over millions of years with addition of new centres for new functions in every new species.

Evolutionary biologists argued then, that man is not an ape like chimpanzee but why not accept that man is an animal who has acquired many more capabilities and intelligence than that of an ape in the process of evolution. We have also a vertebral column supporting our biped erect posture. We have mammalian features similar to apes; we have a liver, heart, lungs and many other organs like apes having more or less similar functions. We also have a brain, though of a bigger size with more advanced capabilities. We have got speech and language which is the privilege of *Homo sapiens* and above that we have consciousness about our abilities. Though our brain might have similar external appearance like that of chimpanzee, chimpanzee can think of building a nest only on tree but man has stepped on the moon and is exploring universe.

Who were our ancestors? What did they look like? What were their activities? Whether they had a language? It is very difficult to find answers to these questions by direct evidence. We have to put together this jigsaw puzzle with the help of fossils of our ancestors, which are parts of their bony skeletons and the stone tools prepared by them scattered all around. Could they have language from the birth? Or was this a gift from God just as Plato and Descartes believed?

Evolutionary Biologists

Evolutionary biologists were still believing Darwin's theory of evolution and its principle that changes occurred in new species because of variations. They could not accept that something can evolve out of nothing. Ancient Greek philosophers as well as modern thinkers, have professed the same. Shakespeare has said 'nothing comes out of nothing'. The great epics of Aryans 'Upanishads' have said 'purnat purnam udichyate' which means 'zero comes out of zero'. It is not that it is just there from the beginning of the universe. As we have seen in the previous chapter on evolution that there are always some variations in the species. Those variations which are beneficial for the survival and reproduction are retained and carried forward. New circuits are added to the brain and new species are evolved.

From the scales on the body, feathers were evolved to maintain the body temperature, while in birds body temperature was maintained by circulating blood, the feathers had no role to play in maintaining body temperature, they therefore metamorphosed into wings which were useful for flying. This is known as the "tinkering effect".

Tinker is a hawker and moves around in the city repairing old utensils. In 1940s, when Dalda ghee (a margarine) was introduced in our kitchens, lot of empty tins of Dalda were available. Many tinkers started moving around in the town and approaching housewives for metamorphosing these tins into useful items. A lid converted it into a beautiful container, a wire handle converted into a pot for planting a Tulshi plant, which is supposed to be a sacred plant and a must, for every household. A hole at the bottom made it a fire place burning 'saw dust', as during war time kerosene was in shortage. A little change here, a little there, a lid, or a wire handle and the same tin was converted into another useful article.

In our villages, women, artfully convert old clothing into beautiful bed covers, 'godhadi', as it is called. They recycled cloth, from different garments and gave it a new function. Magsaysay awardee Nilima Mishra has given an export business of godhadis, or quilt to these ladies. Something like that happens in the living species also. This is known as 'phase translation.' When such changes alter the external look of the species a new form of life—a new species is evolved. Something similar must have happened in the brain of *Homo sapiens* which gave them the art of speech and language. The organs which were originally evolved for swallowing, chewing and breathing were given the additional function of speech. This is known as articulatory system which helps the modern man to speak. Let us therefore have a look at the speech system of modern man.

Speech System of Modern Human

The common man may feel that he speaks with his mouth and a few better informed, might even elaborate more and say that it is the Adam's apple that God has given us, which is why we speak! The Adam's apple is obvious as it projects in the neck, especially in men, and is called the larynx. Both the answers are partially correct. While we speak, we can see only lips and jaw moving and some critical observer may also note the movement of voice box or larynx (Fig. 3.1) moving up and down. But this is not all that is involved in the production of speech. Speech is much more complex art and involves many more organs such as lips, teeth, palate, uvula, and most important, the tongue. The respiratory tract below supplies the air under pressure for the utterance. The rate at which these parts are moving during speech is also amazing. After knowing about it further we will appreciate why speech is known as hidden miracle. And also know why people before Darwin era were saying that this miracle is a gift of God.

The basic unit of language is called a phoneme, the vowels and consonants are also phonemes. English language has only

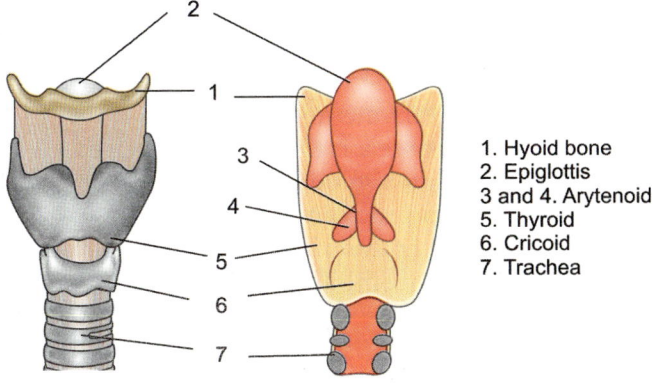

1. Hyoid bone
2. Epiglottis
3 and 4. Arytenoid
5. Thyroid
6. Cricoid
7. Trachea

Fig. 3.1: Voice box from front and behind

40 phonemes, while Devanagari script which is basic for many Indian languages has got 46. An individual phoneme has got no meaning. When vowels and consonants come together they form a syllable. A syllable is uttered in one breath, or word which is a combination of one or more syllables, a syllable and words have meaning. If we wish to tell someone to go away we will utter [g] and [o] in sequence to form a syllable go, which has got a meaning, whereas [g] and [o] individually have no meaning.

How exactly do we speak? The vowels and consonants are each produced by a unique combination of movements that modulate the flow of air coming in SVT or supraventricular vocal tract into unique acoustic pattern. Each vowel and consonant will have a different acoustic character. The total number of muscles which execute these movements of speech organs inclusive of chest, larynx, throat, mouth, face, etc. is about forty. Not all the muscles take part in the production of each phoneme but at least 15 muscles must be taking part in each activity. We utter and understands about 15 to 25 phonemes in one second. This would mean that about 225 or more activities occur per second. About 15 to 20 muscles acting for each of 15 to 25 phonemes per second. We can utter more but then we will not be able to differentiate between the two phonemes due to constraints of our auditory (hearing) system. When someone starts speaking faster we may not understand him. The speed at which we normally speak is faster than any other activity we perform by any other system in the body. If someone starts speaking at a speed slower than that, say at 7 to 8 phonemes per second, that will be also difficult to understand due to constraints of our short-term memory. At that speed our short-term memory may loose the beginning of a long sentence by the time it is finished. Between the two limits is the normal speed or a good speech. Articulation of each phoneme involves the movements of the lower jaw, lips, tongue, lower teeth, soft palate. Daniel Ling has therefore called it a ballets dance of speech. The key player is of course the tongue. The speech far exceeds any other activity in the animal kingdom in complexity. In Farsi language the word for language is 'jaban' which means tongue. Perhaps we were knowing, since a long time that the chief mover in speech is the tongue. This understanding has given birth to expressions like, 'hold your tongue' or 'mother tongue'.

How was this miracle bestowed on man? It is explained by gradation that is, in Darwin's much quoted term 'descent by modification'. To understand the miracle of speech, we need to understand the physiology of sound and speech. Physiology is the science that deals with the function of a biological system. We will now go into the anatomical and physiological basis of the biological systems involved in the production of speech.

Articulatory System

Articulatory system involved in the production of speech consists of three components. The larynx or voice box, which is prominently seen projecting in the neck in males moving up and down, with speech, can be taken as a reference point. The system is divided in three parts: Larynx with its glottal opening, the sub-glottal part mainly consisting of respiratory system and part above the glottis the supraventricular vocal tract (SVT) extending from glottis to lips. The SVT is present to its full length in man only.

Let us start our study of the articulatory system from larynx or voice box as it is also known. Larynx is prominent structure seen in the neck specially in males. The width of larynx from front to back is more in males than females. Larynx is present in all terrestrial animals and is evolved from the valve that protected the lungs of primitive lungfish, while swallowing.

Larynx is made up of 4 soft cartilages (*see* Figs 3.1 and 3.2).

There are two paired and two single cartilages in its structure. The main cartilage named thyroid cartilage, derives the name from tyros (Fig. 3.1), which means "shield". It has two square shield-shaped halves united in front by meeting each other. On the back side they are apart from each other, and are united by a membrane which is fibrous in nature. A single leaf-shaped cartilage called epiglottis, is attached to the notch formed by the two haves of the thyroid cartilage meeting each other and is projected backward. It closes the laryngeal opening by bending itself down, during swallowing. Cricoid is a cartilage having the shape of a wedding ring with its diamond on the back side (Figs 3.2a and b), this is the only complete ring-shaped cartilage of the respiratory tract. It is just below the thyroid cartilage and is attached to it by fibrous membrane. On both sides of diamond two pyramid-shaped small

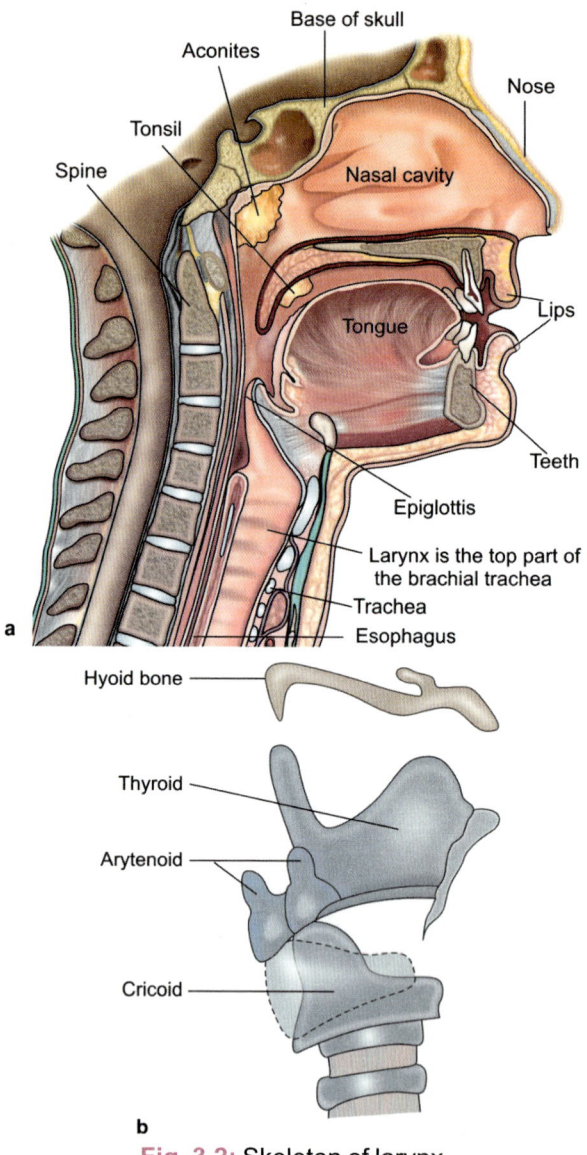

Fig. 3.2: Skeleton of larynx

cartilages are fixed with their triangular bases downwards. As all the cartilages are fixed to each other by membranes and muscles which induce movements of the cartilages in respect with each other. The larynx thus becomes a closed box. There are two cords extending from the thyroid notch to the base of the arytenoid cartilages behind. Though these are erroneously called cords, they are really folds of membrane folded on itself containing a few muscle fibers, some elastic fibers and a few fat globules. As they take part in phonation we will also call them as vocal cords (Fig. 3.3). The opening between the cords is called as glottal opening or glottis. Air can go in and out through this opening. Fresh air with high oxygen content goes in, and air polluted with carbon dioxide goes out. Below the cricoid cartilage, is continuous with the wind pipe called trachea. The cricoid and trachea are fixed to each other by a membrane. All the cartilages of the larynx are connected with each other by the fibrous membrane, open and close glottis (Fig. 3.3).

The trachea descends down in the chest and divides into two main bronchi; the left and the right going to the left and right lung respectively (Fig. 3.4).

The trachea and the bronchi are made up of c-shaped cartilages connected to each other by a membranous tissue.

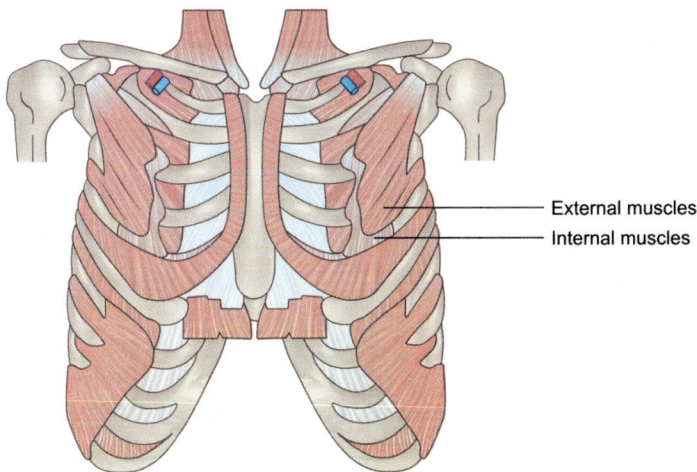

External muscles
Internal muscles

Fig. 3.3: Chest

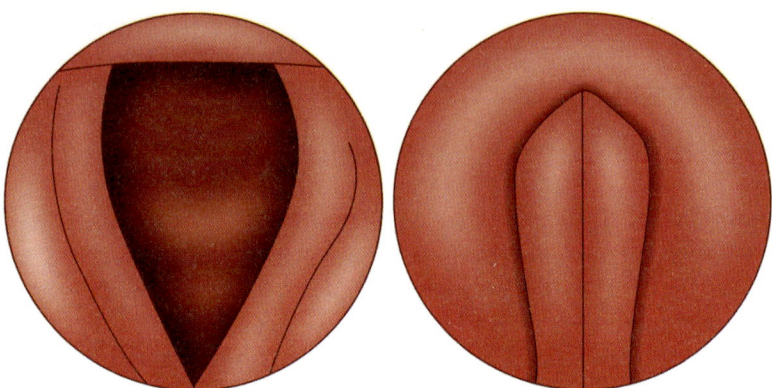

Fig. 3.4: Glottis open and close position

The c-shaped cartilages prevent the trachea and bronchi from collapsing by pressure. Bronchi divide further many times and in the end the smallest has got no cartilage but at its end has a balloon like sack having a very thin wall. The sack is made up of elastic tissue, these sacs inflate with the fresh air coming in and deflate by elastic recoil to drive the polluted air out.

The chest is a closed cavity made up of 12 pairs of ribs which are fixed behind to the vertebral column and in front the chest bone (Fig. 3.4). They are almost part of a circle in shape and the gap in between is closed by two sets of muscle fibers. The internal intercostal muscle fibers and the external intercostal muscle fibers. Attached to the lower margins of lower ribs and vertebral column behind, is a dome-like muscle called diaphragm. Diaphragm separates the chest cavity from the abdominal cavity below. When internal intercostal muscles contract the ribs move in and pressure is created inside the chest cavity. Due to this pressure the air in the lungs is driven out. This activity is called 'expiration' which is also assisted by elastic tissue of the lungs. After the air inside is driven out the external intercostal muscles start contracting and pull the ribs out and up. The chest cavity expands, and relative to the external atmospheric pressure, a vacuum and low pressure area is created inside and the fresh air from outside is sucked inside. This phase is called 'inspiration.' When the external muscles relax, the internal muscles start contracting and bring pressure inside the chest cavity and therefore on the lungs and expel the air out. The inspiration and expiration together form the respiratory activity. It takes place about 16 to 18 times per minute.

Respiration during Speech

Act of respiration during articulation of speech takes place in a little different manner. The air is inspired in the lungs as usual, however, the expiration is lengthened. The vocal cords which had been pulled away to allow the air to come in through the glottis start moving in and the airflow is obstructed below the cords. The external intercostal muscles which are expiratory muscles contract very slowly and air pressure is built up below the cords. When the air pressure is enough to part the cords away the cords move out. And a gush of air rushes out under pressure. The pressure below the cords is thus reduced and the cords start moving in and close again. The cords are brought in because of negative pressure created by the air rushing out of the glottal opening under pressure. This is similar to the little cyclone we observe in the summer on the open grounds. The air is rushing up in circles, and due to its upward force a negative pressure is created around, which pulls up all the things around like leaves, papers, etc. and throw them up. The particular phenomenon is called Bernoulli's phenomenon. The process is repeated till the air pressure built up below the cords is over. Thus, the air rushes out of the glottal opening in puffs of air. **The rate at which the cords move in and out in one second is called the fundamental frequency or 'fo' of phonation.** Frequency is how many times an event repeats in one second? Therefore the frequency at which the cords move in and out in one second decides the frequency of the sound energy available for phonation. To some extent the frequency also depends upon the length of the cord, the tension in the cord and the mass of the cords. The total mass of the vocal cords is decided by the few muscle fibers, and the fat globules in it. The length of the cords in both males and females is almost same till puberty. After puberty the larynx of the male grows longer from front to back. As the length of the cords changes so does the fundamental frequency. In case of males the length of the cord is longer and therefore the fo is less. The frequency is inversely proportional to the length of the cord. It is about 125 to 250 cycles per sec or Hz per sec. Hz is a unit assigned to frequency in memory of Professor Hertz. In case of females as the cords are shorter in length the fo is more and is about 175 to 300 Hz, in case of infants as the length is still less, the fo is still more and is up to 350 Hz. The air rushes out from the glottis at the time of each cycle till air below is over. We normally built up a pressure equal to about 10 cm of water column each time, we speak. This amount of pressure is sufficient to talk for about 40 secs. But if one wants to speak a long sentence or if a singer wants to take a melody of long duration he can built up a pressure of 70 to 100 cm of water column. The loudness of the voice depends upon how far the cords are pushed out by the air and therefore on the pressure built up below the cords. The air puff we expel at one time is sufficient to utter a phoneme. We learn by practice how much pressure will be required to utter a particular phoneme. In case of infants they are still learning this art. Therefore they may speak slowly, or may make mistakes in uttering the particular phoneme or may try to speak fast and finish the talk about which they are not confident. We have therefore, to decide beforehand as to how long a sentence one is going to speak or how long a melody the singer wants to sing at one time.

After Renaissance

French scientist Ferrein (1741) first attempted to explain how the vocal cords produced phonation and published the first theory, that attempted to account, how phonation occurred? Ferrein thought that cords acted like strings of violin. His theory was not totally correct, as we have seen that puffs of air decided the frequency of phonation. But the term cords remained in the scientific world. The cords actually are not like strings, when looked from above they look more like a fold of membranous sheet folded on itself.

Supraventricular Vocal Tract (SVT)

SVT is that part of the articulatory system which is above the glottal opening and extends up to lips above.

Figure 3.5 shows SVT is a tube having variable width and configuration at places. It consists of many organs such as tongue, teeth, lips, etc. The fo is common frequency of phonation and is a common source of sound. When we utter different phonemes (vowels and consonants) each phoneme has got its own frequency. Therefore the common phonetary sound has to change its frequency matching to each phoneme. How this is achieved? Is an interesting story which goes as follows?

The Czar of Russia announced that 'The Academy of science of St. Petersburg Russia', would give a prize to a scientist who could explain the physiological difference between the vowels of Russian language. Kratzenstein accepted this challenge. He prepared a set of tubes which were similar to the shape of human vocal tract. The tubes were fitting in each other. On one side was fixed a vibrating reed just like a flute or shahanai. The vibrating reeds produced a phonetary sound like fundamental frequency. He, by sliding one tube over the other changing the 'effective length' or in other terms 'changing the fo' by extending or reducing the effective length of the tube which was matching to produce different vowels of Russian, and thus he demonstrated the physical difference in the frequency of each vowel of Russian.

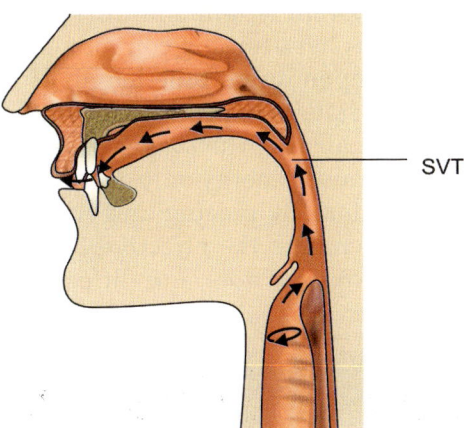

Fig. 3.5: Supraventricular vocal tract

This way he explained the physiological difference that differentiated vowels of Russian language. It goes without saying that he was honored and received the academic prize for the year 1780. The relationship of the production of speech sounds and the SVT is like a flute or an organ pipe. The role of SVT was demonstrated, in this respect, by German scientist Kratzenstein in 1780.

Is not it surprising that man had discovered a flute about 36,000 years ago? It was found in a cave in south Germany, the cave was at Geissenklösterle. The flute was made from a bone of a big bird. This is a long tube narrow at one end and 4 holes were carved on it at different distances (Fig. 3.6). The holes were carved intentionally was proved by the cornices carved around the holes for the fingers to fit snugly on the holes. But to understand that our speech is based on the same principle, mankind had to wait till 1780 when Kratzenstein produced the sounds of Russian vowels, from 36,000 years ago when the first flute was found at Geissenklösterle.

Our speech is based on the same principle of a flute or a pipe organ. The pipe organ has got a hollow bag containing air under pressure and the air is pushed from it by pressing the bag by arm, just as air energy comes out under pressure from our glottal opening. The organ to produce different notes, the air is passed through different pipes attached to it. The pipes are of different length and thickness. These pipes act as acoustic filters between the common source of sound and final output. Depending on the length of the pipes and thickness, different notes are produced. Flute works on the same principle. The air that the player blows from one end acts as a source of acoustic energy or fo as it is called in relation to our speech function. By placing and lifting fingers from different holes we change the effective length of the flute which gives out the desired note, and different notes are thus played.

Fig. 3.6: The flute

In case of man the common source is the air under pressure or energy released through the glottal opening having a fundamental frequency. The interposition of SVT in between the source and the final outcome of a phonemic sound by placing a filter in between, which changes the fo to desired note having appropriate formant frequency. "Formant frequency" is the term for the frequency of the desired note, or vowel or consonant, whatever is the final phonetary outcome. This theory is known as 'source filter theory of speech' and it was originally proposed by Johannes Muller (1848).

Each phoneme has got its own frequency. let us take the example of [a] as in 'about.' It has got its own frequency of 600 Hz.

The filter in the SVT can be likened to the common sieve used in the kitchen. Which sieve to be used depends upon the fineness of the flour required for different needs, depending on recipe to be prepared. For a different recipe a different sieve is used. This can be more clear from an example from our every day life. Our mother in the house uses a sieve to sieve wheat flour. If she wants to prepare a birthday cake for the little son, she will take very fine sieve to have very fine flour. If she wants to prepare some other recipe she will take the other one with a little larger holes. The sieve performs the filter function for the material to be sieved. The matter in case of speech is, the phonetic energy in the form of fundamental frequency of the glottis. The same way a filter is introduced in SVT to change the length of the SVT, which in turn, should change the fundamental frequency or fo. The fo is changed to match the frequency of the phoneme to be uttered. This frequency is known as formant frequency or F1. The formant frequency will therefore depend upon the obstruction acting as a filter in the SVT.

What is acting as a sieve or an obstruction in the SVT? Naturally the tongue, which is the key player in the ballet dance of the organs in the mouth playing for the function of the speech (Fig. 3.7). For example, if we wish to utter the phoneme [i] as in 'ease' the tip of the tongue will be raised and will be almost touching the front end of the palate. When the tongue is removed the air gushes out and phoneme [i] is uttered. The formant frequency of [i] will depend upon the place of the filter and the part of SVT in front of it. As the part of SVT in front of the filter is very small the wavelength of the sound wave will be also small and the frequency will be high (Fig. 3.7). The formant frequency of vowel [i] is 370 Hz. There are of course (Fig. 3.7) higher harmonics such as 3200 called F2, etc. Harmonics higher than that may be there but they are of no significance due to constraints of our auditory system. Our auditory system cannot discriminate between 2 adjacent notes at higher frequencies. Let us take another example.

If we wish to say [u] the tongue will be going up to the back end of the palate and the SVT in front of the tongue will be long. The vowel that will be uttered will have a wavelength long and therefore frequency low (Fig. 3.8). If we wish to say vowel [a] as in "about", the tongue lies flat on the floor, but is raised in the center to some extent but does not cause complete

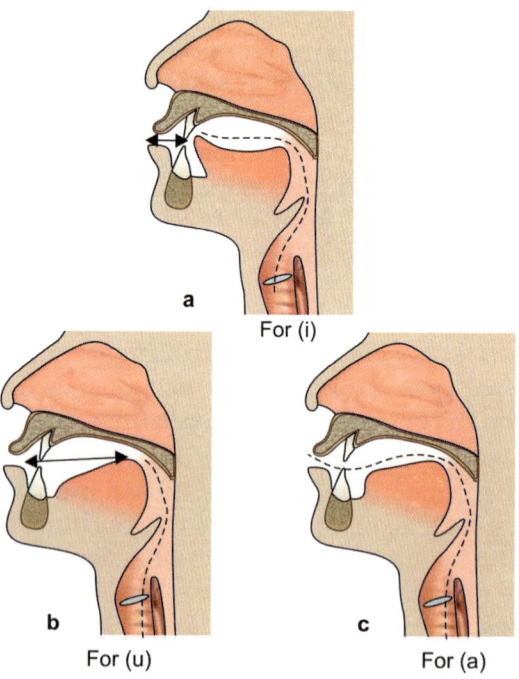

Fig. 3.7: SVT formant frequency of vowel

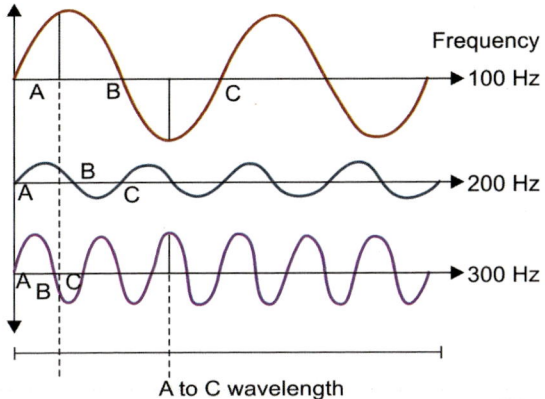

Fig. 3.8: Long wavelength and low frequency

obstruction. The pharyngeal wall at the time of production of vowel 'a' contracts to the extent of 5 mm. Perhaps this is the reason why the tongue is a little raised in the middle. The tongue does not cause complete obstruction. In this case, therefore the whole SVT acts as a filter. The length of the SVT is approximately 17 cm, and therefore, its natural frequency is 600 Hz. The formant frequency of [a] is therefore 600 Hz (i.e. F1). Its harmonics will be 1680 Hz, and above. The higher harmonics are designated as F2, F3 and so on. The formant frequency is the characteristic frequency of the phoneme also called the center frequency for that phoneme. The SVT acting as a complex filter lets maximum sound energy to pass through at that particular level. Just as there are some big holes and some small in a sieve and that will allow some larger and smaller particles to pass through, A band of frequencies also passes on with the **formant frequency. The range of this band is 60 to 100 Hz. Each phoneme has got its own frequency. Let us take the example of [a] as in 'about'.** It has got its own frequency of 600 Hz.

Same is about different musical notes in musical scale. Each note has its own frequency, which has a particular pleasing effect on the human ear. Each such frequency is called a note. This is known as Pythagorean octave. The following table will give an idea about notes in Indian and Western music and their specific frequencies of the first octave starting from 256 Hz.

Symbol of Notation

Western	C	D	A	B	E	F	G	C
Indian	SA	RE	Ga	Ma	P	Dh	Ni	Sa
Western	Do	Re	Mi	Fa	So	La	T	Do
Frequency	256	288	320	341 1/3	384	420 2/3	480	512

Octave is a group of frequencies or notes in rising scale. There are 3 or 4 octaves usually sung in vocal music. Each starting from double the first frequency of previous octave and increasing in the same proportion in the next octave.

The sound energy supplied by the glottal opening of the larynx is of a particular frequency. It is different in male, females and infants. We have already seen that the SVT is divided by the filter or the obstruction. Formant frequency is the product of the part of the SVT in front of the position of the filter. All other frequencies are absorbed. When the obstruction or the filter moves away the glottal frequency however continues from the glottal opening and mixes with the frontal frequency and therefore some peaks and troughs are formed (beats). The position of the beats depends upon the fo or fundamental frequency. Beats do not change the phoneme uttered but it may show some variation in the character of the phoneme. We will see more about it in discussion of prosody.

No sooner is the phoneme uttered the tongue shifts to another position to act as a new filter for the next phoneme (Fig. 3.9).

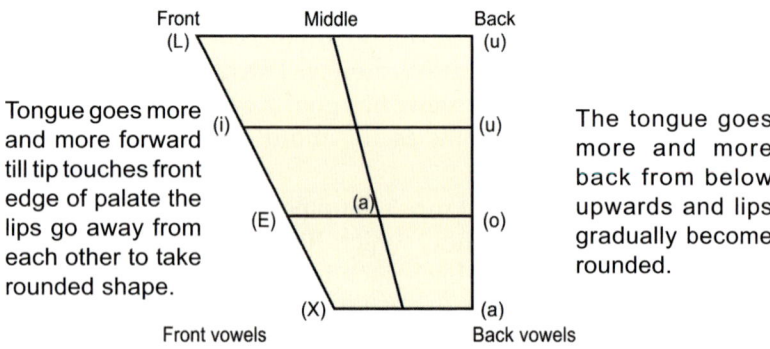

Tongue goes more and more forward till tip touches front edge of palate the lips go away from each other to take rounded shape.

The tongue goes more and more back from below upwards and lips gradually become rounded.

Fig. 3.9: Daniel Jones quadrilateral

For the utterance of vowels the tongue does not actually touch the palate, the significance of this we will be discussing in Chapter 10. For the utterance of consonants, however, the tongue actually touches the palate at the specific positions. Speaker's nasal passage is closed by raising the soft palate at the time of uttering oral sounds. The symbols affixed to the each phoneme such as a, b, c, etc. are empirical and assigned by the society. When a listener listens to a formant frequency 600 Hz he understands that he has heard [a] as in "about" or when he hears a frequency of 370 Hz he knows that he has heard the vowel [i] and so on. These symbols are accepted by the society, and therefore they have entered in the particular language. The symbols are arbitrary. The formant frequency, however, is nonarbitrary.

The utterance of the vowel depends upon the movement of the tongue in respect of the palate. Those vowels which are uttered by tongue going near the front part of the palate are known as front vowels and those when the tongue is near the middle of the palate are middle vowels and those which are uttered when the tongue is near the back part of the palate are known as back vowels. Vowels [i] and [e] are produced when tongue almost touches the palate in front of the middle and therefore are called front vowels, [a] is middle vowel and [u] as in "boot", [a] as in "art", [au] as in "cow" and [u] as in "who" are all back vowels. Daniel Jones represented them on a quadrilateral and has made it easy to understand. It is known as "Daniel Jones quadrilateral" (Fig. 3.9). At the time of uttering of front vowels the lips are wide open and away from each other, while uttering back vowels they are rounded.

We can easily appreciate from the above description of production of vowels that when we utter front vowels the part of the SVT in front of the position of the tongue is small in length. The wavelength of the formant frequency for [i] will be also smallest and the frequency will be high. While at the time of uttering the vowel [u] or back vowels the part of the SVT in front of the position of the tongue is long, the wavelength of the formant frequency will be long and frequency will be less.

This is how we produce different vowels and consonants.

There are seven positions where tongue can make contact with the palate including those for [i] and [u]. These positions are called quantal positions. The tongue cannot go beyond the front and back end of the palate therefore the extreme two vowels [i] and [u] are known as special quantals or steady state vowels. Tongue can make contact at many more positions also but then it will be difficult to differentiate one vowel from the other. These quantal positions also have an acoustic correlate in cochlea and the auditory area in the temporal bone. Therefore tongue having contact with the palate away from the specific position may not give the specific auditory signal assigned for that quantal position. Acoustic correlate in auditory area in the brain is a specific neural property detector. (Property is the particular formant frequency for that phoneme.)

Why Think of Vowels Only?

In all this discussion we thought of vowels only and not of consonants. Why it is so? When the words were formed there was consideration of essentially vowels. The fruit of the neem tree is called 'nimboni' or by some as 'limboni in Marathi language. Both are correct, the meaning of the Latin word 'consonant' means one which comes with. Meaning thereby the one which comes with vowel or consonant. One of my teacher could not speak the 'phoneme 'la'. He had no difficulty at all as there is no 'la' in Sanskrit script. Some adults cannot utter a phoneme say 'ra' but they can carry on without any practical difficulty. Famous old time poet Moropant was writing the epic Ramayan without using a 'labial' and he could do it. Although the names of heroes of Ramayan are all having the names with labials such as, Ram, Laxman, Bharat. Someone came to ask him "Pant what are you writing?" He answered without using labial "Raghunath charit lihit ahe" (I am writing the story of Lord Raghunaths). In short in language vowel is more important than the consonant. Our mother while cooking a vegetable puts something like pulses in the recipe. That is like consonant. In Marathi language it is called 'vyanjan' which means consonant. In culinary skills vegetable is the main ingredient others like condiments, spices are just additions you can do without them, but not without vegetable.

If someone cannot utter vowel say [i] then he will be missing all combinations of consonants with [i]. Same is about [u] and other vowels. In early half of twentieth century the movements of the tongue were still could not be studied therefore the science of articulation was not much advanced.

The discussion up till now is based on our knowledge of acoustic properties of the sound and possible movements of the organs of articulations in the production of a phoneme. As early as 1920 Otto Jesperson postulated that the tongue must be the prime mover, and different phonemic sounds are produced when it touches at different places. It is important to note that Jesperson had no modern testing methods available. It was just his conjecture.

It is simply amazing that 400 years BC the famous grammarian from Indian subcontinent 'Panini' also had this idea. He had explained that the glottal sound source changes in the mouth cavity. He called the glottal source as 'mandra' and the sound quality modified in the mouth as 'madhyam'. He was also aware that there must be some obstruction to the sound source for undergoing this change and also knew that this obstruction was not complete for vowels (Panini's Shiksha). Still we know very little of how exactly different muscles contract and execute these changes. How the characteristics of the vocal cords change for the production of different frequencies? The different areas in the brain responsible for speech were also unknown. In last 5 or 6 decades of 20th century we have made rapid progress in investigative modalities, due to inventions such as cineradiography, FMRI, CT scan, PET scan, etc. And therefore we have come to know more about this hidden miracle.

As a matter of fact all the anatomical parts which take part in the speech production were not evolved for that purpose. They had evolved over many million years ago. The valve which evolved in the lung fish, to protect the respiratory tract from foreign material, later evolved into larynx which plays a very important role in speech production. All terrestrial animals have a larynx (except birds, who have got another organ called syrinx for singing and protecting the lungs). They all inspire air with oxygen and exhale air polluted with carbon dioxide. Tongue, teeth palate, lips are all evolved for cutting food in pieces, chewing and swallowing and drinking. Modifying the use of these parts to produce speech, other than their native function, is a miraculous metamorphosis. The concurrently developing neural control systems were influenced by all these changes. This is known as "Descent with modification" as described by Charles Darwin.

Up till now we have seen in detail how formant frequencies related to each phoneme are related to the position of filter applied to the sound source coming from glottis in SVT. When the phonemes are uttered in sequence one after the other they form a morpheme or a syllable. Morpheme or syllables form words. When words come one after the other in syntactic arrangements, they form a sentence and then semantics. This is how we speak and are understood.

At this point it will be worthwhile spending a few minutes to understand the auditory apparatus or the organs of hearing.

ANATOMY OF THE EAR

When someone speaks, the sound waves travelling through the media reach our ear. Our ear is divided in three parts (Fig. 3.10). The external ear or the auricle we can see fixed on both sides of our skull. The auricles are like horns. Which collect the sound waves. At the center of the auricle there is a depression called concha or a well, which leads to a canal which is about 1 and 1/4th inch in length and is closed by a membrane, at the inner end. This is called tympanic membrane. Sound vibrations entering in the canal sets it in vibrations. Beyond the tympanic membrane there is a small narrow cavity in the bone, the middle ear. The middle ear is a closed cavity lined by bone from all sides except a few openings. The opening on the outer side is closed by tympanic membrane as we have already seen. On the front or anterior wall of the cavity a small opening is seen, which leads to a canal that opens just behind the nose in the nasopharynx. Through this canal, external air can enter in the middle ear and equalize pressure on both sides of the tympanic membrane, to allow it to vibrate freely. On the inner

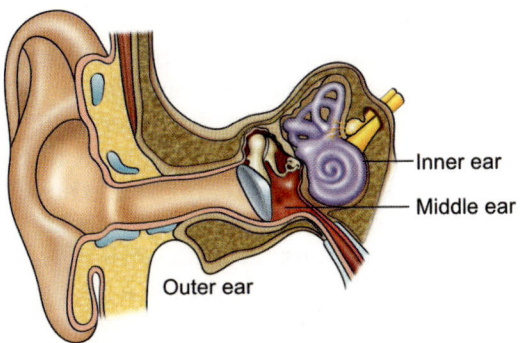

Inner ear

Middle ear

Outer ear

Fig. 3.10: Cross-section of the ear

or medial wall of the middle ear cavity there are two small holes, the upper one is oval in shape and the lower one is round. Both are covered by fibrous membranes. The middle ear cavity consists of a chain of three tiny bones. They are fixed to each other by joint and thus form a chain of bones. The outermost is called malleus; it is fixed to the tympanic membrane. The middle one is called incus and the inner or last in the chain is called stapes due to its shape like a stirrup. It has got a foot plate which fixes on the membrane closing the oval window. When the tympanic membrane is set in vibrations by the sound waves entering in the external canal it sets the chain of bones in vibration.

There is shell-like cavity inside the inner ear, filled with a fluid called perilymph (Fig. 3.11).

Stapes the bone in the chain sets the membrane on the oval window in vibrations. These vibrations are transferred to the fluid in the inner ear via the membrane on the oval window. Here it will be beneficial to go into a little more details of this end organ of hearing. Though the reader might find it a little more in small details it will help us in understanding how we listen the words from sequence of individual phonemes reaching our ear? Inner ear is an irregularly winding bony cavity. The part just behind the oval window is a little widened. The part of the bony cavity behind it is concerned with maintaining the balance of the body. The part in front is called cochlea and is the end organ of hearing. The bony cavity here takes the shape of a shell of a snail. The winding bony cavity is called a labyrinth and is filled with a fluid. There is a membranous sack floating inside the bony cavity which is also filled with a fluid. The inner ear is connected to the cavity of the skull by a canal arising from the inner wall of the bony labyrinth. From the margin of this canal a bony pillar or modiolus enters the shell like cavity in front. A small projection of bone projects inside the membranous tube inside the winding canal. The whole arrangement is like a screw. From the end of this bony partition a fibrous membrane extends to the outer wall of the tube and divides the tube into two compartments. On each fiber of the membrane there is a cluster of cells. This is called an 'organ of Corti' after its investigator. It consists of two pillar-like cells with base on the fiber and upper ends meeting each other up. On the inner side of the pillar there is a row of inner hair cell from below upwards on each fiber. On the outer side there are three or four rows of similar cells. These are called hair cells as there are hair-like projections from the upper side of the cell which are embedded in a jelly-like membrane floating above. On both side there are number of supporting cells. There are nerve fibers attached to the lower side of each hair cell. Figure 3.12 shows on each fiber of the fibrous basal membrane, an organ of Corti is situated thus there are around 3600 organs of Corti responding to different

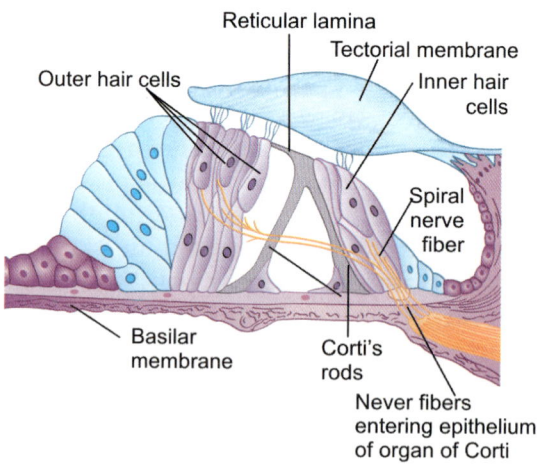

Fig. 3.11: Section of inner ear

frequencies from below upwards in the tube winding round the modiolus. The higher frequencies are placed at the bottom of the tube while low are placed high up. The hair calls are supported on both sides by a few supporting cells. When the sound vibrations reach the inner ear via stapes congruent vibration in the fluid inside and the basilar membrane are set in. The cells on the membrane receive the vibrations and there is movement of electrons inside and outside of the cell, which induces an electric wave congruent in all respects to acoustic or sound wave. This electric wave is carried by the nerve fibers attached to the lower end of the cell ending in the cavity inside the modiolus. Fibers coming from all the organs of Corti inside form a bundle which is the 8th nerve and it enters the skull cavity through the internal auditory canal and then enters the substance of brainstem. From there on it travels a long distance in the brain to reach the auditory center in the temporal lobe.

In the brain the electric current is decoded into speech. The electrical waves reaching the auditory center are exactly congruent to the sound waves incident on the ear in all characteristics. How do we perceive sound from the electrical waves we will learn in Chapter 10.

This system is species specific. Even in animals similar system works. In case of frog there is a small cluster of cells acting as organ of Corti. When a male frog gives a mating call, the female listening to it responds. There are 130 species of frogs, but the female of a species will respond to the call of a male belonging only to the same species. The calls are thus species specific (Goldstein 1963 and Capranica 1963).

Brain and its Function in Speech Production

We have already seen how a particular phoneme is uttered. This is only external expression by the articulatory system. Before a phoneme is uttered it has to be organized in the brain and order is to be given by the brain for its expression. This was a mystery only a few decades ago. Since the new techniques such as PET, FMRI, etc. came in, we are able to peep into the functioning of the live brain and have acquired more knowledge to solve this mystery. For organizing this complex language and ordering for its utterance one will naturally feel that the whole brain must be working. But it is not so. With these new techniques at hand we can find out more or less exactly where the different functions are executed. To understand that, we will take a look of the different parts of the brain.

Anatomy of Brain

The brain is kept secured from the outside impacts, inside the skull. *Homo sapiens* brain weighs about 1200 to 1500 grams. When you open the skull cap the brain looks like an opened walnut. There are elevations and depressions on the surface called the gyri and sulci, respectively. The external 6 to 7 mm part of the brain looks grey in color and is made up of neural cells (Fig. 3.12). They are about 100 billions or so in number. The inner part is white in color and is made up of fibers arising from these cells (Fig. 3.13).

The neural fibers arising from the cells are of two types. One is a long one and is called axon. The other ones are short and thick and more in numbers, they are called dendrites. The axons take away the messages from the neural cell to another cell or other parts of the body. The dendrites are incoming fibers and bring messages from other cells or fiber complexes towards the neural cell. Each cell keeps contact with about 10,000 or

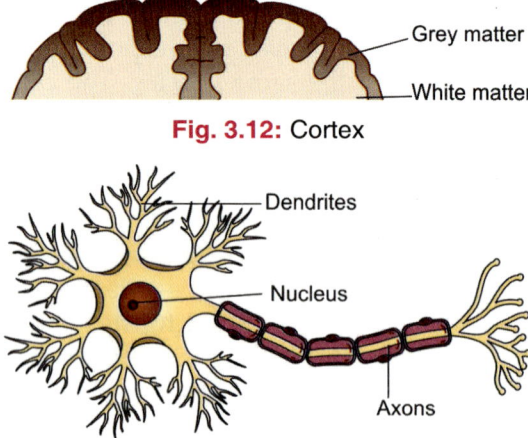

Grey matter

White matter

Fig. 3.12: Cortex

Dendrites

Nucleus

Axons

Fig. 3.13: Neuron

about 10,000 or more cells. A complex network is formed in the brain of cells and fibers coming in or going out. There are some islands of cells in the substance of the brain performing a specific function. A complex formed by a cell, axon and dendron coming out of it is called, a neural circuit. There are millions of such circuits in the brain. These neural circuits are designated for a specific function. If you open a TV set or a radio from behind you will see a network of wires going from one center to another. There are many such centers. They perform different functions such as an amplifier to boost the current, a transformer for changing the voltage as needed, etc. There are such clusters of neural circuits performing different functions which can be compared to such an arrangement. Similarly there are centers in the brain performing different functions such as an auditory center for hearing, a visual center for vision, and so on. The stimulus or the message travels from one center to another through a neural circuit in the form of an electric current of very low voltage which can be tapped externally. The situation of these centers can be fairly accurately located by FMRI. Say for an example auditory center. When we hear, the auditory center functions more vigorously and consumes more oxygen which is seen as more activity in the center in FMRI. Similarly centers for different functions can be identified. Brain is functioning all the 24 hours. Continuously stimuli are traveling from one center to another. From these stimuli we are all the time receiving knowledge of the environment and our position in it.

Brain and its Parts

After opening the skull the brain appears like a fruit of walnut for which shell is removed. On the surface there are sulci or fissures and on their sides there are elevations or gyri. From their location or the particular function they perform they are named differently. The brain is divided into two halves by a deep central sulcus or fissure going from front to back dividing it into the right and left hemispheres (Fig. 3.14). The deep fissure is known as longitudinal fissure. Cortical part of the brain is recently evolved in last few millions of years and as such is known as new brain. The pre-human species also posses new

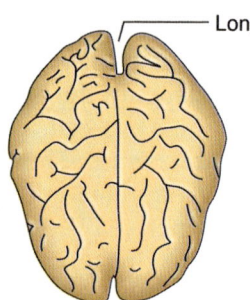

Longitudinal fissure

Fig. 3.14: Brain cortex

brain. In reptile class there is no new brain. The old brain they are having is called reptile brain. As the new species evolved and acquired new functions, new neural circuits develop and the size of the brain slowly increased.

Humans are classified in the class 'vertebrata'. The vertebrates are having a vertebral column on the back made up of several individual vertebrae. At the center of the vertebral column there is a canal like space going from skull downwards (Fig. 3.15) through this canal runs the spinal cord, from the brain high up, to the lower end of the canal. The spinal cord is made up of several nerve fibers bringing in sensations from the various parts of the body, and the fibers going out. The fibers going out are the motor fibers reaching to the various muscle groups of the body for executing various movements. High up the spinal cord enters the skull cavity through a large opening called 'foramen magnum'. As it enters, it expands and is then called 'medulla oblongata', being oblong in shape. This is the lowest part of the brain. The medulla oblongata consists of clusters of neural cells, called nuclei controlling various vital activities of the body such as control of beating heart, respiratory activity, etc. Therefore injury to this part may cause sudden death. It also contains the first nucleus of auditory sensation, sense of balance, movements of the face, eyes, etc. The part above is still enlarged and rounded and is called midbrain. Behind the midbrain and medulla is the cerebellum which is also a part of reptile brain. Cerebellum is concerned with keeping the balance of the body in biped erect posture. It is also a center for appreciating rhythm. Above the midbrain

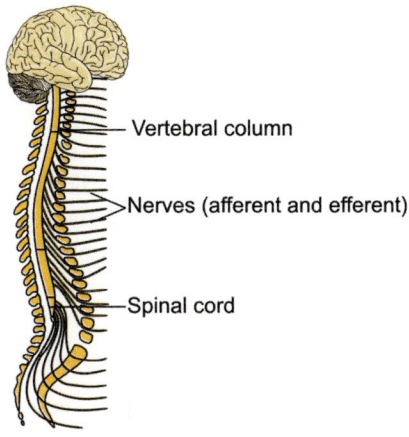

Fig. 3.15: Vertebral column and spinal cord

there is a part like a platform and it serves really as a platform for the new brain consisting of two hemispheres (Fig. 3.16). The platform consists of thalamus—the meaning of this Latin word is platform—hypothalamus and basal nuclei. This part performs all the important functions in reptiles. The functions of this part we will know in more details later. Just above the platform there is a bridge like part which joins the two hemispheres, known as corpus callosum.

Both the hemispheres of the new brain or cerebrum are similar in external appearance. However, different functions are carried out in hemispheres of both the sides. From about the center of the longitudinal fissure, another fissure starts and

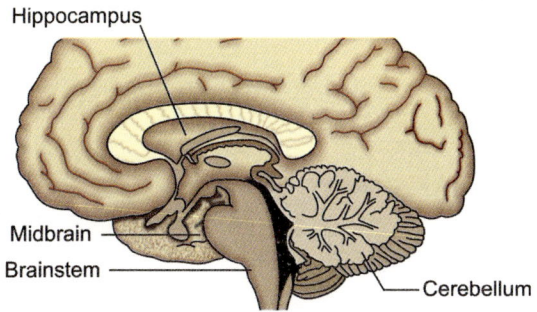

Fig. 3.16: Basal nuclei

passes downwards and sidewards (Fig. 3.17). It divides the brain in two parts. This fissure is called central fissure. Part in front of this fissure is called frontal lobe and the part behind is parietal lobe. The elevation or the gyrus in front of the central fissure in the frontal lobe is the motor area, it controls the muscles of the movements of the body. The parts of the body are represented upside down on this gyrus (Fig. 3.18). The part controlling the feet, thighs, abdomen, chest, head, neck, and

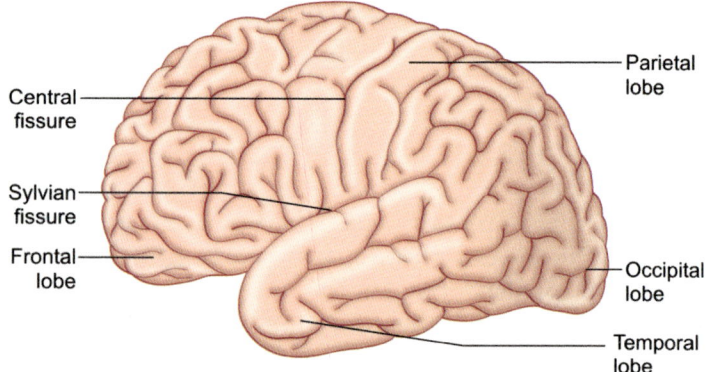

Fig. 3.17: Lateral view of the brain

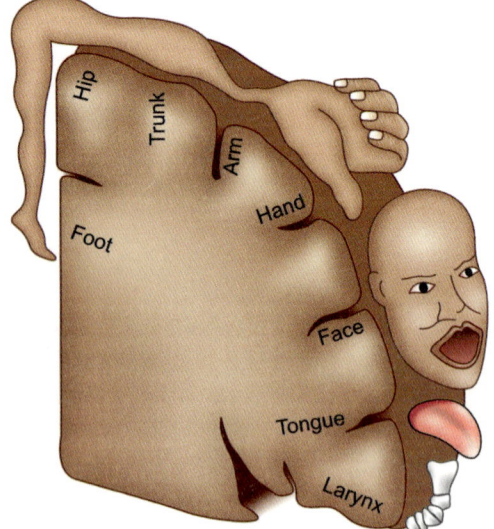

Fig. 3.18: Homunculus

face are represented in that order from above downwards. The elevation behind the fissure receives the sensations from the body such as pain, touch, temperature, joint sense, etc. They are also represented in the same fashion feet and legs above and head, neck and face below. Just in front of the area controlling the motor activities of the mouth, lips, tongue and jaw, there is a little elevated area in the frontal lobe, called Broca's area (Fig. 3.19), which controls the articulatory movements of the speech. We will get to it in more details in Chapter 10. The parietal lobe performs many functions in modern human and therefore is much enlarged.

From the lower end of the central fissure one more deep fissure starts which runs upwards and backwards. This is a 'sylvian fissure'. The part of the brain behind the sylvian fissure is known as 'temporal lobe'. Behind the center of the fissure and on the gyrus behind, is the auditory area. The sound waves received by the ear ultimately reach this area. Majority of the nerve fibers coming from the right side reach the left side and vice versa. Just behind and above the auditory area is another area which is known as Wernicke's area after the German scientist Wernicke who first reported about the function of this area (Fig. 3.19). The main function of the Wernicke's area is

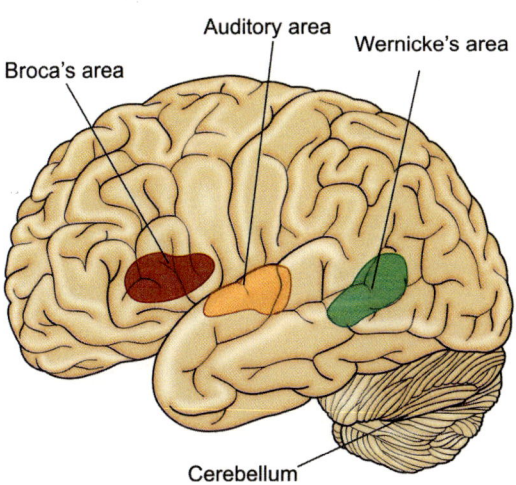

Fig. 3.19: Different centers in brain

comprehension of the stimuli received by the auditory area. The Wernicke's area is 7 times larger in modern human than in chimpanzee. The sylvian fissure is longer and deeper on the left side of the brain. Wernicke's area organizes the matter to be uttered by the articulatory system. This area also composes complex sentences. Wernicke's area and Broca's area are both connected by a bundle of neural fibers from below the sylvian fissure. This bundle of fibers is known as arcuate fibers (Fig. 3.20). Through this channel the speech organized and composed by Wernicke's area is carried to the Broca's area for utterance. On the inner aspect of the temporal lobe there is a part somewhat looking like a hippo and that is why it is called 'hippocampus area'. It is more or less semicircular in shape and at its anterior or front end is a globular area known as amygdala (Fig. 3.21). Hippocampus area stores the long-term memories. The amygdala is concerned with primary emotions. It receives as well as initiates emotions and recognizing faces. This way the functions of receiving and comprehending auditory information, organizing the speech matter, composing complex sentences, preserving memories, recognizing faces, and emotions are all functions undertaken by the temporal lobe. The complex consisting of hippocampus, amygdala, cingulate gyrus, basal nuclei is known as limbic system.

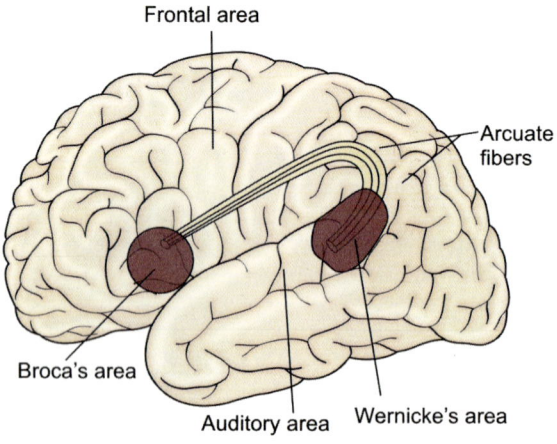

Fig. 3.20: Arcuate fibers

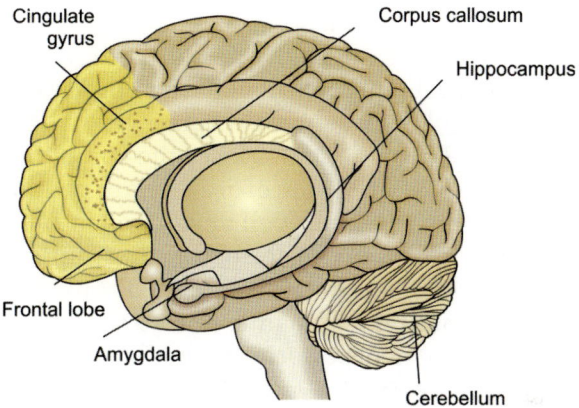

Fig. 3.21: Limbic system

When we are in dialog with someone the words that we listen are stored for short time in short-term memory area which is called 'working memory'. After the sentence is over if necessary we extract some words from the word memory bank of the brain and answer. 'Working memory' is really of very short duration lasting only till the sentence is over. If there is a telephone call in between or we have to take down a telephone number, we lose the reference and have to ask the partner to repeat. This happens even when he speaks in long sentences, here we may lose the beginning, by the time we reach the end of a long sentence.

Broca's Area

Paul Broca, French scientist first reported the importance of this area in articulation of speech. One of his patient had an accident and head injury. He had lost his speech. Unfortunately the patient did not survive. After postmortem examination, Broca found out the lesion was in frontal lobe just in front of the area of the gyrus which supplies the motor fibers to the face, lips, tongue, etc. After this initial observation of Broca, similar findings were detected in many other cases, not only of accidental deaths but also of left hemispheric lesions due to other reasons, leading to right side paralysis. The area is known as Broca's area in memory of Paul Broca. In this area neural circuits are formed for the articulation of different phonemes, which is the smallest part of the speech and language. Broca's area also performs the function of composing the matter to be uttered, in syntax (sentence). The matter organized by the Wernicke's area and sent for articulation to Broca's, is uttered in unorganized rubbish manner when there is damaged to the Broca's area. The patient also loses the conjunctions such as, (and, or) he cannot therefore organize compound sentences. The Broca's area is well developed in *Homo sapiens* and creates an impression on the skull cap from inside. The Broca's area on left side is mainly related with speech, the same area on the right side is mainly related with music, poetic expression, etc. A patient suffering from right-sided paralysis due to lesion in Broca's area on the left side is asked to recite a poem, he can do it fairly well. The area of the frontal lobe just in front of the Broca's area is related with abstract thinking.

Basal Nuclei

Language is not just uttering of phonemic sounds in the form of words in sequence. We utter the words in different loudness, give stress on some parts of the words, take a pause at certain intervals, and lengthen some words. All these characters of the language are very important. They stress the importance of some part of our speech or even change the meaning and intention of our speech. These are known as suprasegmental and prosodic aspects of the speech. The pattern of fundamental frequency may play a role in signifying the end of the sentence. This is same in almost all the languages of the world and is called as linguistic universal. Stress on some phonemes may change the meaning of the word. For example let us look at the word permit, if a stress is given on the phoneme [e] pe'rmit it becomes a noun but if stress is given on phoneme [i] permi't it becomes a verb. Several such examples may be found in any language. These are called suprasegmental and prosodic aspects and are common to all languages. 'You had been to a movie yesterday? can be a statement, if stress is given on the word movie 'you had been to the movie?' It becomes a question. The same conversation between father and son in a serious tone, putting an exclamation mark in the end, would have a completely different meaning? These suprasegmental and

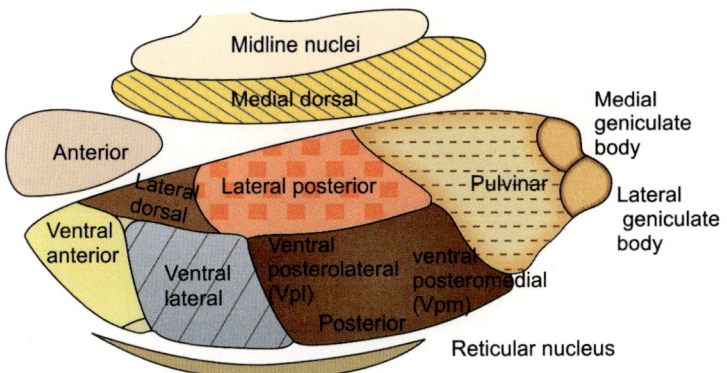

Fig. 3.22: Basal nuclei

prosodic functions are related to thalamus and basal nuclei. Just as in an office, copies of letters are sent to different departments. Similarly, copies of the message sent to articulatory system are sent to basal nuclei, thalamus and hypothalamus (Fig. 3.22).

They add the necessary changes as per requirement of the speech and send it further for utterance. This process is common to all the languages. In diseases of basal nuclei like Parkinson's disease speech becomes monotonous without any prosodic effects as there is deficiency of dopamine which is required to stimulate the function of basal nuclei. Function of the frontal lobe is to maintain the self of the person, i.e. intelligence and cognitive ability. There are special neurons known as 'mirror neurons' at some special places in the brain. We will be discussing more about mirror neurons further in Chapter 9. The hindmost part of the brain is called an occipital lobe. It is concerned with the sense of vision. There are about 30 sub-centers where comprehension of the sense of vision takes place. One of that is angular gyrus (Fig. 3.23).

The most backward and lowermost part of the parietal lobe is called inferior parietal lobule or IPL. In IPL the upper part of the Wernicke's area, angular gyrus comprehending visual sensation, the sensation of the joints and position of the muscles, from that part of parietal lobe behind the post central gyrus, all are coming together. In case of modern human the IPL area

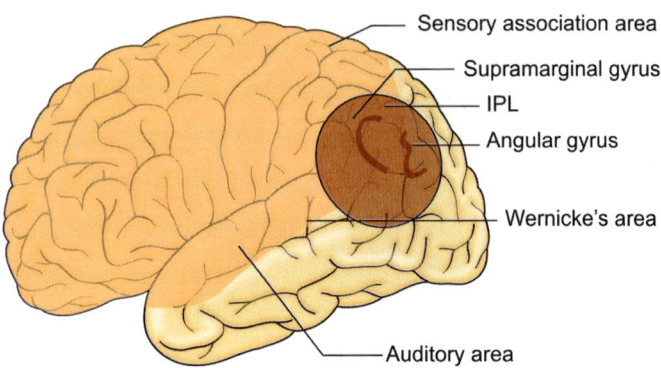

Fig. 3.23: Inferior parietal lobule

is much enlarged. Because of this enlargement the lower part splits into two separate gyri. One of the two is known as 'supramarginal gyrus and the other one is angular gyrus. The supramarginal gyrus is concerned with delicate movements of the fingers for delicate works such threading a needle holding a branch firmly, etc. It is also concerned with subtle/movements of the tongue, face, etc. to show subtle emotions and to prepare delicate weapons for better use. Thus, this area where the visual perception, perception of auditory sense, and knowledge of position of joints and body come together is very important. We will also call it IPL. The IPL plays a very important function in the production of speech. Mirror neurons are in abundance in this area.

Up till now we have taken the relevant knowledge of the different parts of the brain and their functions related to different aspects of language.

Let us take an overview **in the form of a chart** (Table 3.1).

We have learnt up till now how vowels and consonants are uttered in a sequential manner to form the words. However, we have not still answered the two questions asked by Tinbergen. Who composes and organizes the material to be expressed and who orders the composed matter to be uttered?

Before answering these deep rooted questions, we will have to answer a few problems which though allied are not trivial.

Table 3.1: Parts of the brain and its function and importance

S. No.	Parts of brain	Function and importance
1.	Upper 5 to 6 mm grey and below, white	Gray consists of neural cells, and white of fibers, axons and dendrites
2.	Neural circuits, cells and the fibers arising from them	
3.	Brainstem	The part connecting the brain and midbrain controls all the important activities of the body such as respiration, heart beats, and control on movements of face, tongue, larynx and eyes
4.	Cerebellum	Rhythmic activities of body keeping the balance of the body in erect position
5.	Frontal lobe	Maintaining self of the individual, intelligence and cognitive knowledge and control of motor activities of the body
6.	Broca's area	Control on articulation of speech, composing sentences, and use of conjunctions
7.	Parietal lobe	Receiving sensations from different part of the body, there are plenty of mirror neurons, in lowermost and most backward part or IPL vision, auditory and touch and joint sense all come together
8.	Gyrus in front of central fissure	Order the motor activities of the body
9.	Gyrus behind the central fissure	Receiving sensory stimuli from the body
10.	Temporal lobe	Auditory area, auditory comprehension, composing the matter to be delivered

(Contd.)

Table 3.1: Parts of the brain and its function and importance (*Contd.*)

S. No.	Parts of brain	Function and importance
11.	Wernicke's area	Comprehension of received auditory stimulus, compose semantics, and send the message to Broca's area
12.	Arcuate fibers	Taking the stimulus from the Wernicke's area to Broca's area
13.	Hippocampus	Store long-term memory
14.	Amygdala	Center for emotions

When the modern man started speaking?

Was he the first amongst pre-human species to have this skill of oral speech or was any previous species speaking? If not why?

Just because of anatomical constraints or there was neural immaturity also.

Second important question is where exactly the *Homo sapiens* uttered the first word? One view is *Homo sapiens* evolved in Africa and before dispersing out of Africa they had already acquired language. The other view is there was multi regional dispersal of *Homo erectus* a pre-human species and *Homo sapiens* evolved from them wherever they migrated. This was easy way to answer the question why there are many languages in the world. After we discuss these issues we will answer the two basic questions raised by Tinbergen and then we will solve the puzzle how man acquired language.

Up till now we have talked about speech, but what about music? Was music already there? Or both developed in tandem or music was just a lucky breakthrough of language? Let us have a look to that important part of language, the music in the next chapter.

Music and Language

Language is not a single package containing only statements, but it can be expressed in musical form too. Even if we accept language evolved by natural selection, question remains how music was evolved? Whether music was there before language and did pre-human species use it for communication, entertainment and acquisition of cognitive knowledge and culture. At what place in the brain is music appreciated? Music is of two kinds, one is pure music or classical music and instrumental music, and the other one comes with lyrics. Where are the centers for both? What is the relation between the music and emotions? How the music stimulates the emotions? Has music got some other effects on human psyche? Can music be used for treatment of ailments? What is the function of community music? We will try to find out answers to all these questions in this chapter.

The Dnyaneshwari holds a special place in the hearts of every Maharashtrian, if not every Indian. It is the insightful commentary on the Bhagavad-Gita (the foundation of Hindu religion.) written by the Marathi saint and poet Dnyaneshwari during the 13th century at the tender age of 16.

While the Bhagavad-Gita is a unique statement of spiritual knowledge professed by Lord Krishna to Arjun, the Dnyaneshwari is the more easy, expanded version told with lucid examples about the teachings of Bhagavad-Gita, aimed at the behavioral development of the common people.

In this work written in 1300 AD. The very first verse begins with an invocation to "Aum". Why the Aum?

Hindus believe that as creation began, the divine all—encompassing consciousness took the form of the first and the original vibration, manifesting as the sound "Aum". Thus, it was considered by them to be the original unique sound.

A reflection of the absolute reality "Aum" is said to be "aadi anant", i.e. without beginning or the end, and embracing all that exists. It is also supposed to be the divine energy (Shakti) united in three elementary aspects, creation, preservation and destruction.

Of the three sounds comprising "Aum" the a: Akaara implies form or shape the u: Ukaara implies formlessness and or/ shapelessness (e.g. that of fire or water) and the m: Makaara implies neither of the other two, but something that still exists, e.g. darkness.

Thus, by invoking the "Aum", saint Dnyaneshwari had perhaps clearly implied that this musical vibration is all encompassing and existing even before the life on earth began. In this context music seems to have originated before language.

Bruno Nettle distinguished ethno-musicologist and professor emeritus of Music and Anthropology has defined music as 'communication outside the scope of language'. Greek philosophers of ancient period had called it as "music of Spheres" perhaps similar concept as "Aum."

MUSIC: AN INNATE ABILITY

Even a lay person can appreciate performance of an Indian classical singer without having learnt it. Appreciation of music has known, no boundaries be it Mozart, or Bach or Bhimsen Joshi or Pandit Ravi Shankar. Most people, be he, a surgeon or a car driver enjoy music playing in the background, even though they themselves may not sing or play an instrument.

In his well translated work "How Musical is Man" John Blacking (1973) has put forth his belief that inherent musical ability is the defining characteristic of being human. It is an innate ability that need not be learnt, only tapped into and refined with more inputs.

It is, therefore, fair to say that awareness of knowledge of music precedes that of any language, and every human being has an intuitive and in born sense and knowledge about music. This is best honed before it gets muddled with the left brained logical worldly activities. Without formal training individual from any culture has the ability to appreciate music in some fashion to make it , why so? Is it because it is in our biology? May be. This might be the very reason, why early music lessons (started at 3 years of age) are shown to be extremely effective in learning music?

LANGUAGE AND MAN

In 1866, a conference on "The Science of Language" was held in Paris, but the 'society Linguistique De Paris' refused to admit any papers on the origin of language. There was not any mention of music in connection to language.

Precious little has been said about where or in which part of the brain is music understood or sensed.

Charles Darwin has briefly touched upon it in his work "descent of Man" and has been supported by Kenneth Miller. Darwin's studies suggest that music originated as a tool to attract the female of the species for the purpose of mating, copulation and survival of the species. He claims that from the survival view point, music was not required in day to day business. Before language got developed, musical sounds with different pitches and tonalities was the only way to communicate with or attract the female.

As Steven Pinker puts it while language has long been considered essential to unlocking the mechanisms of human intelligence, music is generally treated as evolutionary frippery an auditory cheesecake.

The concept of signaling with sounds is a trait that can be seen even today in human infants, who are completely dependent on their parents for survival. While communicating with babies, the pitch, tonality and rhythm are found to be of supreme importance, for a baby will often burst into peals of laughter at even nonsense uttered in a familiar musical tone.

Ian Cross, a musicologist from Cambridge University supports the idea that an infant is more attentive to sounds and notes and that this has been the case in our predecessors (the *Homo erectus* and *Homo neanderthalesis*).

Scientists have always been intrigued by the connection between language and music. Decade long neuroscientific research suggest that language and music both contribute in telling us, who we are and from where we have come from? Jackendoff and Lerdhl investigated language and music as cognitive phenomena. Breaking music into its components, rhythm, the structure of melody, harmony and emotion in music, they tried to find out which elements of music arise from general cognitive process, which are common to music and language and if anything is peculiar to music. The way people convert music into gesture whether by dance, in conducting an orchestra is instinctive and special to music.

Is the music entirely abstracted from the world we live in? Schwartz and colleagues do not agree, they say "explanation of music like the explanation of any product of mind, must be rooted in biology. Music is rooted in our environment, like visual art of the natural world. It emulates our sound environment in the way visual arts emulate the visual environment. In music we hear the echo of our basic music making instrument the vocal tract. The explanation of human music is simpler still than Pythagoras's mathematical equations. We like the sound that we are familiar to, specifically we like sounds that reminds of us.

The special pitch receptors to detect, a musical note say a (F sharp); regardless of the instrument it is played on, are relatively a few in number in case of human. While in case of a Marmoset (an ape with a bushy tale) they are located in a small area of its auditory cortex and embedded with a much larger number of units. These units respond to low frequency sound in the range of 125–2000 Hz, and are mainly active in right brain via direct circuits or through thalamocortical circuits connecting basal nuclei to cortex.

However, once the language gets developed, it gets precedence as a tool of communication. Yet, even today, we find that the tonality, pitch or stress given on a particular word, or even parts of it, impact the meaning of the communication. Similarly, piece of music may invoke different emotions among different people. These reach right brain cortex connecting basal nuclei to cortex via thalamocortical circuits, whereas words of a language which reach direct to left cortex provide a greater consensus on meaning.

Different people have put forth divergent views on the subject:

1. Elizabeth Talbert (Music and Meaning, an evolutionary story) feels that development of music and language must have been concurrent.

2. Robin Dunbar in his book Titled "Primate Societies, 1988" claims that language got developed through music.

3. A lot of people even today, feel that music got developed as a spin off, while language was evolving. In other words, many believe that music originated as source of entertainment. (Steven Pinker) and is therefore an alternative form of language communication.

In the case of a hearing impaired child who is deprived of hearing sounds and therefore associating them with emotions, it has been observed that expression of emotions becomes very difficult. Very often, this might lead to bottling up of emotions and/or emotional outbursts and aggressions. An early intervention that can improve hearing and communication at a very early age has been seen to dramatically improve the emotional wellbeing of a hearing impaired child.

Difference between Music and Language

Literature/books in one language can be translated into another, e.g. English to French. Translation of music is not so simple for the following reasons:

a. Western music is a common denominator across huge geographical areas and regions, e.g. most of the developed countries predominantly follow a particular type of music, and hence a question of translation is somewhat redundant.

b. However, conversion of certain type of music (say Indian classical music) into some other type (say Western classical) will perhaps not be possible in totality due to the wide divergence in the systems followed in notes and notations as well as the instruments used.

c. Music is timeless as compared to language. Bach, Beethoven continue to inspire generations. Also, there is an universality in the response to music, e.g. foot tapping, nodding and shaking heads, etc. Words in a language have got a specific meaning associated with them contextually or symbolically, and find a common acceptance in large groups of people in a particular society. Notes or tunes of music on the other hand, may convey different meanings to different people, and it may also depend on their emotional state, their cultural background and environment. That music generates different emotions is a well-proven fact and has been supported by evidence in the form of EEG findings.

d. Grammar syntax and positioning of words is important in language, and interchanging of words may perhaps significantly alter the meaning, interchanging of notes in the music on the other hand might not significantly change emotions conveyed or expressed. Certain form of music are even solely rhythm oriented and words are considered insignificant like the 'taranas' in Indian classical music which comprises words such as 'tana de re na' ... which has got no meaning, but are merely, fit onto the rhythm and tune of the music. Certain other forms such as rap music treat actual musical notes secondary while the focus is on words and rhythm.

e. Language conveys information and is predominantly used for this purpose. To draw a parallel music conveys or impacts on emotional states (and not necessarily information).

A language can be used explicitly to write/talk on a topic, e.g. "The influence of Hollywood on Hindi cinema." Music by itself without using words is incapable of doing so.

Similarities

1. Both find oral expressions, can be expressed through gestures and be written too.
2. A combination of language and music (songs) are considered far more effective than each in isolation.

3. Poetry which can be viewed as language combined with rhythm, that is one aspect of music is also capable of producing a universal magical effect, e.g. see William Wordsworth's famous poem Daffodils.

> "The waves beside them danced, but they
> outdid the sparkling waves in glee,
> A poet could not but be gay
> In such a joyous company
> I gazed and gazed and gazed but little thought
> What wealth the show to me had brought.
> For oft on my couch I lie,
> In vacant or in pensive mood
> They flash upon the inward eye
> Which is the bliss of solitude?
> And then my heart with pleasure fills,
> And dances with the daffodils".

The rhyming pattern and the words invoke universal feelings in the mind of the reader and are successful in almost painting visual picture. Language combined with the elements of music (rhythm, tunes notes) produces an impact which is far greater than if Wordsworth would have used only words and not poetry to express himself.

Impact of Music on Humans

The seventeenth century poet William Congreve claimed that 'Music has charms to soothe a savage beast or to soften rocks, or bend a knotted oak.' Among the extremely famous ancient classical singers in India, Baiju is said to have sung so well that a rock actually broke. The more famous singer Tansen (one of the nine jewels of the emperor Akbar's court) is said to have sung Raga Deepak (a musical composition which is supposed to enlighten the flame within) and indeed the atmosphere became so hot that he fell ill. The story is concluded with two of his disciples immediately singing Raga

Malhar, which actually brought down showers of rain. The story might be a little exaggerated but we can definitely say that music has got an influence on perceived atmosphere to a significant extent.

Music therapy is gaining popularity today as one of the soothing therapy. It is believed to cure or at least provide temporary relief from physical and mental ailments.

Emotions originate in the limbic system and amygdala, which is a part of the limbic system. Plato thinks emotions kill the logical thinking in human. Darwin in his book "Expression of emotions in man and animals" says that emotions are phylogenic remnants of our ancestral instincts. They are of no use to modern human. A lot of thought process has evolved ever since, with special mention of the book by Sloboda and Patrick which significantly explains the connection between emotions and music and also the very important role emotions play in physiology of modern human. Music is one way to express emotions basic as well as more complex ones. It is no wonder that it is seen to be very effectively used as a background score for films and serials. The score of Titanic or any love story as well as poignant notes of Dr Zivago certainly produce emotions that the director of the film would have wanted to create in the minds of the audience.

Most of the stories of our evolutionary processes are not to be seen so much as replacing old capabilities of ancestral brain for new one, or rather extending those by embedding them in an enriched system. They implicated the mirror neuron system in human emotional reactions to music (in emotional states think of facial expressions instantaneously produced) and mirror neuron system can read emotional responses back when we hear music in 2006 (Molnar-Sazakacs 2006).

A person visiting the grand canyon was reported not to get the same effect of the magnanimity of the wonder, as he had earlier felt when seen on television, perhaps he had seen it on TV with effective background music (Anthony Storrs 'Music and the Mind').

Baby Talk or Motherese

Baby talk or what is called "Infant Directed Speech or IDS" by Dissnayake is very important in early life of an infant before it acquires language competency. Every mother knows that an infant shows more interest in rhythm, tone and melody than a talk. At this stage of infancy IDS is the best way of communicating with an infant. Highly exaggerated prosodic elements accompany this talk. IDS is an universal phenomenon. Be it mother from a very high society, a village woman or a mother from a tribal group from deep forest of Africa. Each mother knows that an infant shows more interest in a talk with musical score.

A fetus starts having auditory experiences from 5th month of fetal life. It has no language competency at that stage. At that stage infant shows more interest in singing by the mother than her talk. It is also observed that an infant shows more interest in the song, which it had heard in fetal life, and is smoothened more quickly by the same tunes. Ancient Indian medicine has given more importance to keeping a good linguistic environment around an expected mother. Not to have harsh words or occasions of quarrel which will disturb the mother.

Jayne Stanley has shown that music was found to stabilize oxygen saturation level which enhances the physical development of premature infant. Premature infants subjected to a combination of good care and music were fit for discharge on an average eleven days earlier than a control group.

Community Singing and Dancing

Community or folk music of tribes, village population, folklore and traditional music of ladies has seeped slowly in urban population to. It has got special significance. McNeal has studied the benefit of community music and dancing amongst Kalahari Bushmen in Africa. They were convinced that it gives a sense of togetherness to them, and relieves from stress of daily routine. The evolutionary psychologist Robin Dunbar has proposed that any physical activity, specially like community dancing leads to surge of an enzyme 'endorphin' within the brain tissue. This induces feeling of wellbeing. It has also got pain relieving effect.

Community singing and dancing develops a sense of social bonding amongst the group which is very essential in the life in the deep forest infested with predators.

Place of Music and Language in Brain

The key question we will have to answer is, if music evolved earlier than language and whether the same neural circuits earlier used by music were latter used by language? Or whether these circuits were used by something else earlier and latter borrowed (exaptation) by music and language. The other possibility is neural circuits for music and languages are entirely different. If the neural circuits used by music and language are the same when one is lost the other also will be lost, but if the neural circuits used by both are different then when one is lost other will still remain intact. In technical terms this is called as 'double dissociation'. We will have to answer one more question where these neural circuits for music and language are placed in the brain. How can we find it out? There are three ways of doing it. One is to take help of modern investigative techniques; the other is studying some cases suffering from disease and third is by postmortem study if unfortunately the patient succumbs to the ailment. Let us start with study of some medical cases of those who had by trauma or disease lost the speech.

VY Shebalin was musician of 20th century. He was a professor at Moscow conservatory and was teacher to many composers. He suffered from a mild stroke when he was 51. He had right-sided paralysis and disturbed speech. At the age of 57 he had another attack. At the age 67 he had a sever attack to which he succumbed. His speech was lost during his first attack only. Shebalin, 5 months before his death had composed his 5th symphony, this was considered by Dmitri Shostakovich as a brilliant creative work, filled with highest emotions, optimistic and full of life. Shebalin's case is published in Journal of Neurological Science in 1965. Shebalin continued to compose and teach his pupils by listening to their compositions. His hearing was not affected.

Professor A. Luria and his colleagues from the department of psychology of Moscow university have studied his case in detail. They thought that his case provided an evidence that the brain has quite separate neural systems for music and language.

On September 1980, a 24-year-old woman was admitted in neurological clinic in Strasbourg following sudden difficulty with language, as she was unable to understand others. She was under care of Dr Metz-Lutz and Dr Dahi. She was able to recognize nonverbal sounds, could recognize musical instruments from their tunes, and could understand familiar melodies. The success was better when the melodies were hummed rather than sung with words. Words hindered her recognition. She could understand prosodic effect on the basis of intonation. CT scan examinations one year after the onset of patient's difficulties indicated left temporal hypodensity, and a vascular lesion in the left middle cerebral artery. The processing of intonation, as argued by Dr Metz-Lutz and Dr Dahi takes place in the right hemisphere and was therefore not inhibited. She had difficulty also in understanding foreign language. This was because foreign languages may have relied as much on the typical intonation of the language as on the words themselves, to be able to undertake 'lexical decisions'—that is discriminating between words and non-words.

Both the above cases are of adults. Is it possible that earlier in childhood both the neural circuits of music and language were the same and later they got dissociated from each other? Let us look at the next case.

The child psychologist from Illinois Pro. Miller was called onto examine a case of a 5-year-old boy Eddie. He had visual, hearing and learning difficulties, had hardly any linguistic abilities and was physically handicapped. He was able to play piano surprisingly well. He played the Chrisms carol 'Silent Night' and Miller was impressed. Miller thought that he might have played it by practice. Miller therefor asked him to play 'Twinkle, twinkle little star'. Not only he could play it well but could play it in different pitches. Miller has reported 13 such cases in his book "Musical Savants, Exceptional Skill in Mentally

Retarded." All had perfect pitch, or absolute pitch. Though it is very common to have perfect pitch in very early age hardly one in thousand in adults has it.

Thus, far we have seen cases where capacity of understanding and production of music without words can exist in the absence of language and speech. Can converse be also true? If it is so then we can say that it is the situation of double dissociation. Both the neural circuits are separate and even if one is lost the other can exist. Their evolution and growth also must have taken place separately. Following cases will show that it is so.

Let us look at the case of an Australian 65 years of age. He had an attack of hemiplegia of left side. He had good speech reading and writing. Previous to the attack he was a good lover of music and use to play piano well. After the attack his speech was good and reading and writing was also intact. As the attack was left-sided hemiplegia, his damage was on right side and left-sided Broca's area was intact. He knew what to play but his fingers were unable to move when placed on piano. He disliked music after the attack and felt musical melodies as shouting. On CT scan examination of the brain, it was found that he had lesion in right side IPL. As his right-sided supramarginal gyrus in IPL was damaged his fine movements of fingers were lost and he was unable to play piano. He was unable to play other musical instruments too. This case again proves that centers for appreciation and execution of music are situated on right side of the brain and those of speech and language on left. Thus, there is 'double dissociation' between the two.

One thing must be remembered here, if someone attends the musical score not from the view point of enjoying it but for critical analysis, that takes place on the left side of the brain and he will not enjoy the music. As, that time his left brain undertaking critical analysis, is actively dominant and right appreciating music is recessive.

Isabelle Peretz from Montreal University is of the opinion that music is a spatial sensation, just like art, painting, etc. Appreciation of rhythm was evolved after the human got bipedal position and heel walk. The rhythm is therefore, appreciated by left brain and cerebellum. Cerebellum is a part of old brain and has got long evolutionary history. When we listen music auditorily the stimulus is divided in various modules processing tone, pitch, melody, etc. and we get a collective appreciation and comprehension. If you ask an arranger of the music which instrument is more important, it will be difficult for him to tell, it is a collective effect that is soothing to the ear. It is not so in case of language, the words come one after the other, and that is known as temporal sensation while music is received by spatial sensation as described above.

Lawrence Parsons of University of Texas undertook an ambitious research program with the help of PET. His findings were, the scales of music and rhythm activated the middle temporal area with greater intensity on the left than right. While symphony of say Bach comprising tunes, tones and melody activated superior, middle and inferior temporal gyri with greater intensity on the right than left.

Just as cerebellum shows activity in appreciation of rhythm, basal nuclei are activated in appreciation of prosodic and suprasegmental factors which are part of musical stimulus. Thus, neural networks for music processing extend beyond the cerebral cortex into the parts which have got a long evolutionary history. This has got phylogenic importance in evolution. In animals who have got no neocortex the vocal calls are given by old brain or reptile brain. The music is however appreciated by right temporal lobe and adjacent area of frontal lobe in higher primates and modern human.

Appreciation of Lyrics

One more aspect of music is appreciation of lyrics as they have got both, meaningful words as well as all the other characteristics of music. There is story of a Canadian, lover of lyrical music KB. He had a huge stalk, tapes and disks of lyrical music. He had an attack of hemiplegia. If a lyric was played with its tunes it was understood. But if the same tunes are played with another lyric he could not understand it. As the tunes were fixed in his memory with the same lyrical words.

Both, words of the lyrics and the tunes have got separate modules or neural circuits which process them. By repeated auditory experience interconnections are fixed between the words of the lyric and the tunes. Therefore when the tunes were played he remembered the lyric which was fixed in his memory with the tunes but when the words are changed he could not understand either words or the tunes.

Music and Hearing

Even if hearing is lost by some ailment the music remains. World famous composer Beethoven had become severely deaf by disease of otosclerosis after his middle age. This is a disease of middle ear and gives conductive hearing loss, the neural circuits in the brain are not damaged. Even after hearing loss, he could compose his famous symphonies.

World famous linguist Noam Chomsky says that just as competence of learning language is innate, the music is also in our biology from the birth. Same is the opinion of Anthony Storr who has expressed it in his book "Music and the Mind".

Conclusion

Thus, music is in biology of all living beings. Jagadish Chandra Bose has undertaken experiments on plants and has stated that plants respond to music. Some of the agriculturists like Shri Nirgudkar have claimed that the music can influence the growth and yields of the plants. Whether we accept these claims or not, but one thing is certain that music is there in our biology and we have not to learn that but we have just to take it out as Blacking has said. How, when and why it is encoded in human genome are the few questions still unanswered? As said in the beginning of this chapter Greek philosophers called it "Music of Spheres", originated with the spinning of universe. Saint Dnyaneshwari has also called it "Aum" the adi or sound from the beginning of the universe.

5 When *Homo sapiens* Spoke First?

When exactly the modern man acquired this miracle of speech? Did any other pre-human species possess this art? And if not why? What anatomical and physiological changes took place which lead the modern human to acquire the speech? Did our cousins Neanderthals also had similar anatomical configuration? Did they have language and if not what was the difference which prevented them from having speech? What are the proofs to show that modern human species only, acquired the speech about 150 thousand years ago.

INTRODUCTION

How were our ancestors living? How they were communicating? Had they any speech? If yes, from which pre-human species was it existing? If no, why they did not have the compositional language? None of the member of the pre-human species is living today, anywhere in the world. None of these questions therefore can be answered today by any direct evidence. We have already seen how modern human can speak. What was special in *Homo sapiens* that they possessed this miracle, the compositional language? Was it only structural (anatomical) competence, which was not present in pre-human species or something more? All these are the questions we have to answer.

We are all familiar with the parents doting on the newborn and fondly capturing the milestones as the baby learns to speak. They are witnessing the evolution or development of language in the child. The following dialogue is commonly heard in a household where there is arrival of a new baby.

"Listen oh dear make a note of today's date in our baby's 'Baby book' she called me, mom today." "Oh what is there to be getting so much exited, she called me, da, da, da yesterday only." "Oh, do not be so jealous, baby always calls her mother first, is not it dear!"

The archeologists, evolutionary biologists, anthropologists, linguists and other scientists on the other hand did not have this luxury. They had to make do with the fossils and imagine and piece together the jigsaw puzzle. They had only fossils of pre-human species and the stone tools prepared by them scattered around as the only evidence available to them. The fossils do not talk and the language does not leave back any fossils. Then when the modern man acquired this miracle of speech or was it present before?

To solve this question is like solving mystery of Agatha Christi or AV Gardener. The key man Hercules Piorot collects small, small evidences, and reaches the conclusion of the puzzle and points to the culprit and declares "yes, you are the murderer." We have already seen that there is no direct evidence left. Under these circumstances the 'circumstantial evidence' is sufficient and accepted by the court to prove the guilt. Perry Mason the lawyer detective of Erle Stanley Gardener says that sometimes circumstantial evidence, if all of it points to only one conclusion, is enough to put the rope around the neck of the killer.

For us there are three factors or evidences to be considered before coming to the conclusion that when a particular species had phonemic or compositional language.

1. Whether the articulatory system was competent to utter phonetic sounds of the language?
2. Whether the evolution of brain was advanced enough to organize the speech and language and order the utterance of it?
3. Language is the key factor in the cultural development of a society. Are the archeological findings, suggestive of such advancements in the period when that society was living on the earth? That will be an indication to whether the language was evolved, in that pre-human community or not?

Fossils found at box grove in south of England belonged to a species which had all the characters which were indicating that this was a species which was showing more signs of evolutionary progress than the species *Homo erectus*. This species was labeled as *Homo heidelbergensis*. Similar find was explored from the Sahara region of Africa. In Europe, a new species evolved from Heidelbergensis about 300, thousands of years BP, which was named as *Homo neanderthalesis*. Another species evolved in Africa about 150 to 200 thousands of years ago from African Heidelbergensis, which was named as *Homo sapiens*—the species to which we modern human belong. We have thus common ancestor with Neanderthals or we can say that Neanderthals are our cousins. If we can convince ourselves that Neanderthals who evolved before us were not competent of uttering phonemic speech, leading to compositional language, we can certainly say that no other previous species had speech and language, and speech evolved only with the evolution of *Homo sapiens*. Let us now explore the evidences to prove our hypothesis.

Anatomical Competence

We have already seen in Chapter 3, how we speak, what organs are involved in speech mechanism and how they work? With the help of modern investigative techniques, such as cineradiography, functional MRI, etc. we can now actually see their movements. From the fossil studies of Neanderthals we will have to find out the anatomical configuration of articulatory organs of Neanderthals, and whether they were competent to utter the phonetary or compositional language. If not capable of it, then why? This is a branch of study known as comparative anatomy. Sir Victor Negus for the first time published a book in 1949 "comparative anatomy of larynx and pharynx."

Speech Anatomy of Neanderthals

What was the speech anatomy of Neanderthals? We have seen that speech anatomy consists of three parts. Part below the larynx or voice box and part above it.

The part below is the respiratory system. The chest of Neanderthals is barrel-shaped as against that of *Homo sapiens*, which is somewhat kidney shaped. The barrel-shaped chest was not perfectly competent for the finest respiratory control required for correct utterance of a phoneme. Phonation is the term used to describe the process by which the larynx produces a periodic series of puffs of air. Monkeys, apes and human have similar larynges. Neanderthals also had the same. The part above the larynx is the supraventricular vocal tract (SVT). Before going into the details of configuration of SVT we will have a look of the general profile of the skull of Neanderthals. It will be interesting to have a simultaneous study of the skull and SVT of a newborn child (Figs 5.1a to c) Neanderthal, and modern human adult. The skull of newborn human child and Neanderthals are flattened from the top to bottom and elongated. Their lower jaws are proportionately longer than those of adult humans, and they have got no chin. The newborn and the Neanderthal faces project in front of forehead the part of the skull that encloses the frontal region of the brain. This projection is one of the characteristics that differentiates Neanderthal from modern human. The skull cap in Neanderthals is not flexed while in human it is flexed. In case of modern human more neural circuits are formed in the frontal lobe. The frontal lobe is the place where the individual thinks of his self, and maintains his individuality. His self-thinking capacity is advanced, he can think of his capabilities and how to use them, his cognitive intelligence is advanced and has acquired cognitive fluidity or interdomain intelligence as Carruthers has called it, which was not with Neanderthals. He has acquired new emotions such as love, arrogance, hate, shame

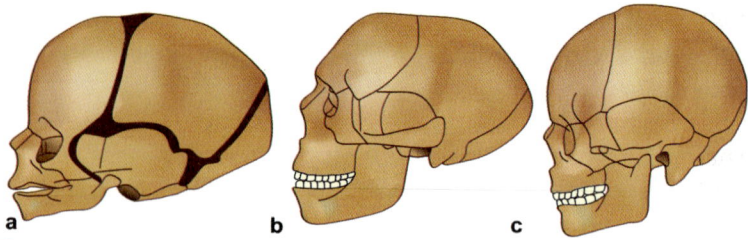

Figs 5.1a to c: Skulls of; (a) infant; (b) Neanderthal; (c) modern human

and many others with his intellectual capacity and social life, thus it can be seen that the frontal lobe has given us our self by which we are the culturally most advanced species. All the centers for all these developments are mostly in forebrain part of frontal lobe. Many new neural circuits must have developed in this region which increased its size and weight. The net result was projection of forebrain and flection of skull, both found only in modern human.

Crelin's reconstruction of vocal tracts of Neanderthals put them parallel to those of newborn human infant but latter studies proved that the vocal tract of newborn is more like a chimpanzee. The larynx is high up just behind the tongue in the mouth and the whole of the tongue is in the mouth. In Neanderthal the larynx has shifted lower down just as in modern human but not to the same extent. Crelin prepared casts of vocal tracts from the fossil of Neanderthal French village La chapelle-aux-sent, of adult human and an infant. The casts of adult human and infant were checked against the X-rays of the living human to assess the correctness of the method and therefore the correctness of the cast of Neanderthals (Fig. 5.2). The view of the bottom of the skulls of adult human, infant, and Neanderthal will make it more clear. In case of adult human the erect biped posture is achieved and the skull is flexed. The reasons are mentioned above. In case of Neanderthal, however, though biped stance is achieved the skull is not flexed and the jaw is longer, the larynx had not fully gone down and therefore the length of the mouth cavity and the pharynx is not equal, the length of the mouth cavity is longer. In such a case the tongue is thinner, long and almost lying in the mouth. As the jaw length is longer the tip of the tongue cannot reach the front end of the palate.

If we note the human newborn skull it looks as if, it was similar to that of big ape or chimpanzee. Ed. Crelin the world's authority on infant anatomy and author of first book on 'Anatomy of newborn infant' in 1969, tried to reconstruct the vocal tracts of Neanderthal and newborn infants and adult human. He took into consideration, the marks and ridges made by muscles attached or firmly glued to the skull base for

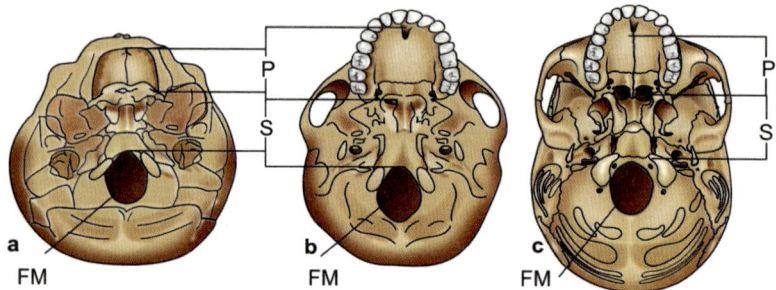

Figs 5.2a and b: Skull bases of (a) newborn infant, (b) Neanderthal, (c) adult human. P: Length of hard palate. S: Distance between back end of hard palate and anterior border of foramen magnum. FM: Foramen magnum. Distance P and S are equal in infant and Neanderthal. In adult human S is less than P

reference. The bottom part of the skull supports the soft tissues of the vocal tract such as lips, tongue, pharynx and so on, with the muscles attached to it. Crelin was able to make a reasonable attempt at reconstructing the Neanderthal vocal tract (Fig. 5.2) shows the bottom view of the skulls of an infant, adult human and Neanderthal. The newborn infant and Neanderthal skull bases show striking similarities. One of the keys to understand how the Neanderthal differs from that of the adult human is the length of the palate, and the distance between the back end of the hard palate and the front margin of the foramen magnum. In Neanderthals they are 6.1 and 6.2 cm respectively, while in adult human they are 5.1 and 4.5 cm they are labeled as P and S respectively in (Fig. 5.2). In an infant the distance between the palate and the foramen magnum is about 2.6 cm and the length of the palate is about the same or a bit longer. The space inside the vocal tract is hard to visualize. But understanding the dimensions of this part is very important to understand the speech capabilities of any one. The correctness of the reconstructed casts of the vocal tracts of Neanderthal, infant and the adult human were confirmed by the X-rays and cineradiography records of newborn and living human adult. In an infant the larynx is positioned just behind the tongue close to the bottom of the skull. After the age of 3 months when the infant lifts the head and soon starts sitting up, the larynx starts its descent down in the neck and reaches its adult position by the age of 7–8 years. In adult human the distance between the palate and the foramen magnum is shorter as mentioned above and cannot accommodate larynx. Neanderthal SVT therefore was neither like that of an infant nor like an adult. In archaic hominids and the Neanderthals the length of the mouth and pharynx were not equal. We have already seen that when the larynx starts shifting down, because of pressure of forward moving vertebral column, to achieve biped erect position it pulled down the tongue also with it, and then the posterior 1/3 part of tongue was now facing backward and a new cavity is created behind the tongue called pharynx. The part of the pharynx above the level of the tongue behind the cavity of mouth is called oropharynx. If we try to accommodate the length of the pharynx equal to the mouth in a Neanderthal vocal tract, the larynx will be pushed down in the chest an improbable proposition (Fig. 5.3), a monster.

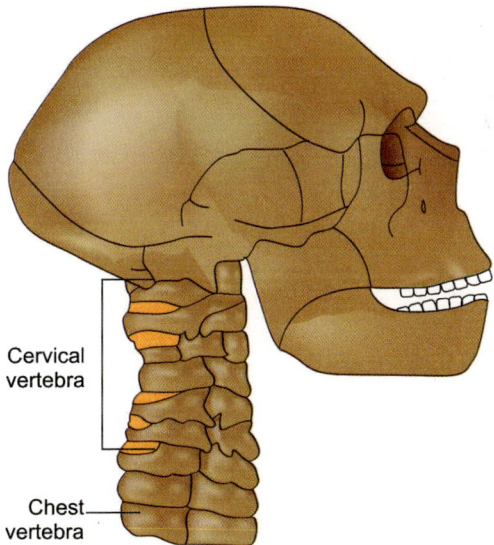

Fig. 5.3: If length of mouth and pharynx are imagined to be equal length larynx may go to chest (a monster)

After the biped position the mouth cavity ends at the back or posterior edge of the palate and a new part pharynx is developed behind. We have already seen that this part of pharynx behind mouth cavity is called oropharynx, it has got no anterior or front wall, it directly opens in the mouth. The pharynx extends down behind the larynx up to the opening of food pipe. The distinctive qualities of the human speech derive from the equal length of mouth and pharynx and both of them being at right angle to each other figures above show the sagittal sections of heads of Neanderthals and modern human (Figs 5.4 and 5.5) show clearly the angle between the mouth and pharynx in case of Neanderthal (Fig. 5.5) below it is not at right angle and therefore the tongue cannot be raised to back end of hard palate while production of back vowels. While in case of modern man (Fig. 5.5) above the mouth and pharynx are at right angle to each other and tongue can be raised and touch the back end of hard palate to procure obstruction required for back vowel.

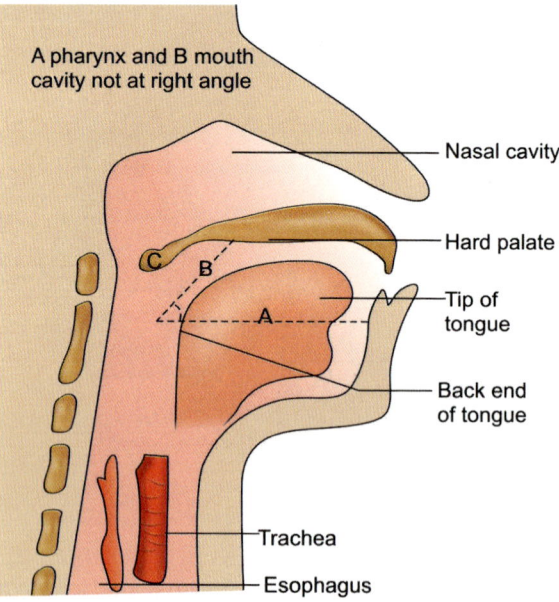

Fig. 5.4: Relation of pharynx and mouth cavity in Neanderthal

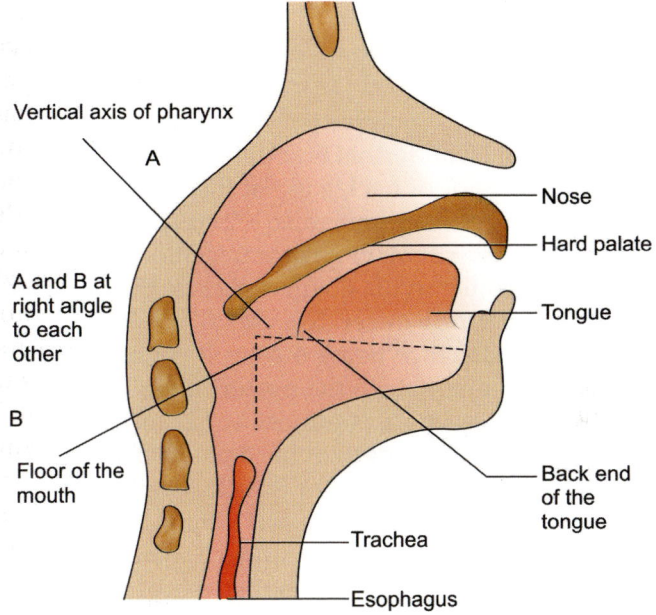

Fig. 5.5: Relation of pharynx and mouth cavity in modern human

The equal lengths of these parts enable us to produce all the formant frequencies specific for various phonemes of the language including the vowels (Fig. 5.4). One more important point is the angle between the hard palate or the length of the mouth cavity and the top of posterior 1/3 of the tongue, which forms the beginning of the laryngopharynx. In case of Neanderthal it is not at right angle, while in case of human it is at right angle. The back part of the tongue is rounded and when it goes up to cause obstruction for the vowel [u] it must almost meet the posterior end of hard palate, as these two points are not at right angle it cannot do so and therefore Neanderthal cannot produce vowel [u] and consonants [k] and [g]. We have already seen that as the mouth cavity of Neanderthal is long and protruding further the tongue tip cannot reach the front end of palate which is the requirement for the production of front vowel [i]. In short the fact remains that Neanderthal cannot utter steady state vowels at both ends, namely [i] and [u]. All the quantal positions on the hard palate are fixed in reference to these two steady state vowels naturally Neanderthals would not be able to utter other vowels also accurately.

One more important point is to be considered. The length of the hard palate of Neanderthal is about 6.2 cm and the distance between the posterior or back edge of the palate and the front edge of the foramen magnum is also 6.2 cm. In short the length of the oropharynx is 6.2 cm while pronunciation of the phoneme the soft palate and the uvula go up and close the upward path. Air therefore cannot enter the nasal cavity. In case of Neanderthal the distance being 6.2 cm it cannot be closed as the tip of uvula cannot reach the posterior wall. The length of the soft palate with uvula is almost equal to hard palate. In case of Neanderthal the length of hard palate being 6.2 cm and the opening of the oropharynx or the path to nose is the same 6.2 cm. The uvula can just reach but cannot effectively close the path to nose while in case of modern man the opening of the path to nose is 4.1 cm and the length of the soft palate complex being equal to hard palate is 5.2 cm. Therefore it can effectively close the opening. In case of Neanderthal therefore as some air can escape to nasal cavity during pronunciation the phoneme uttered has got nasal character. While in case of human it is not the case.

What Neanderthal could not Speak?

What Neanderthals could not speak? It can be easily deducted from the differentiation between the anatomical configuration of the two as discussed in the section above. Neanderthal is unable to produce important front and back steady state vowels [i] and [u] therefore all other vowels will not be having a defined quantal position, net result will be they are unable to produce all the vowels properly. They are unable to produce consonants [k] and [g]. The jaw of Neanderthal is long and relatively the tongue is short, as part of it has already descended down in the pharynx along with the larynx. The tip of the tongue has to reach the front end of the palate to produce the vowel [i] as it is the front vowel. If it tries to reach and produce required formant frequency, as there is a gap between the tip of the tongue and the front end of the palate, palate being longer and the jaw protruding, the air will escape and there will be only a hissing noise. Similar thing happens when the tongue tries to utter the extreme back vowel [u] and the glottal consonants [k] and [g]. As the mouth and pharynx are not at right angle to each other, the back part of the tongue cannot be elevated to reach the required back position. The tongue normally takes a position near the floor of the mouth and its top surface is rounded as the back 1/3 part has gone down. In case of Neanderthal the descent is not complete and therefore the tongue cannot take up the stance required for the production of vowel [u].

The vowels [i], [u] and [a] are very important. Roman Jacobson has demonstrated that in any language, amongst all the vowels, the vowel [i] comes about 93.5% times, vowel [u] at about 88% times and [a] about 83% times. We know the experiment conducted by Gordon Peterson, in which vowel [i] and [u] were missed only 2 to 5 times by 10000 volunteers against all others, which were missed for more than 500 times. We will know more about this work in Chapter 10. One can imagine how many words Neanderthals must be missing and the language if at all they could speak any, must be unintelligible?

There is one more factor that makes the language of Neanderthal unintelligible and that is the speech of Neanderthal will be nasal in character for the reasons mentioned in the previous section. Zivi Bond has undertaken a research in this respect, working with cases of cleft palate and those with palatopharyngeal incompetency and found out that if air escapes from behind the nose the speech will be nasal and such a speech is difficult to understand. The listener will be missing about 50% of the speech. We also get this experience when we hear speech of a child having a cleft palate.

After all this discussion we can be in a position to decide as to what Neanderthal speech means. Vowel production is governed by the position of the tongue, lips and larynx. The correct vowel production depends upon the basic physiological and acoustic constraints. In case of Neanderthals the tongue is already pulled down in the pharynx to certain extent. The jaw is however still longer than that in *Homo sapiens*. There are therefore limits on how much maximum constriction that can be produced between the tongue and the palate without producing turbulent noisy nonvowel air flow.

We have already seen that Neanderthal vocal tract is incapable of production of the universal vowels [i], [u] and [a] of all the human languages. These vowels also delimit the total range of vowels that any normal human can produce. As Neanderthal cannot produce the vowels at the limits of total vowel range, it is doubtful whether they can produce the vowels which happen to have their positions in between them. Moreover all the speech sounds they produced would have had a nasal quality.

Neanderthals were thus inherently unable to produce compositional speech and language. It would have sounded like an idiot incapable of talking the right language.

Development of Neural Structures

The brain size of Neanderthal was as big as or a little larger than *Homo sapiens*. Does that mean that, they were as intelligent as we are? No. Intelligence does not depend, only on the size of the brain. We have already seen that, in our chapter on evolution. We have also seen there the reasons why the size of Neanderthal brain was larger. It is the functional neural anatomy that decides what brain can do? The size of inferior parietal lobule or IPL is increased 7 times in human than that in chimpanzee. This causes an impression on the inside of the skull cap. In Neanderthal there is no such prominent impression. As the size of IPL was not that big. There was no splitting of IPL in, angular and supramarginal gyri. IPL plays a very important part in the evolution of language. But the fact is, there is not one single documented paleo-neurological evidence, that unambiguously demonstrates the relative expansion of IPL junction. Therefore there was no question of any 'transmodal abstraction' between visual sensation and auditory sensation. It can be concluded therefore that Neanderthal were unable to organize and produce phonemic sounds and words as in modern human.

The question is, though the vocal tract of Neanderthal was capable of producing all other vowels of human language except [i] [u] and [a] though less efficiently, whether the neural circuits for the production of the vowels were developed? Though they possessed these capacities their speech was more susceptible for missinterpretation because of the defects that we have mentioned above as Neanderthal were not able to utter them.The quantal positions for all other vowels except [i] and [u] were not very accurate as the positions for these two was also not fix. It is quite likely therefore that they would not have made use of that, in daily life for communication. The nature is amazingly conservative in production of new organs, unless they are beneficial for the species. Therefore as Neanderthal was not using this phonetic language as it was unintelligible, even if they could produce it partially. If they were not making use of it, it is doubtful whether the neural circuits were developed for the production of these vowels. To produce new circuit is costly as more fuel is required to keep them alive.

Mutations and natural selection for rapid appropriate motor responses required for rapid production of phonemic sounds could have contributed to the enlargement of many of the neuro-anatomical structures implicated in human language. Four-fold enlargements of parts of the human brain that are active when we talk or comprehend language could in itself account for the qualitative difference between the human brain and that of primate and also brain of pre-human species including Neanderthal.

Quantitative enlargements of prefrontal cortex, cerebellum and basal ganglia, neuroanatomical structures involved in regulating motor control, syntax and thinking, differentiates the human brain from that of primates and pre-human species. So far as we can tell from brain volume and overall packaging of the brain these changes were in place 100,000 years ago in the hominids who could talk as we do.

Cultural and Social Development

We have seen in the chapter of evolution that the cognitive development of Neanderthals was of single domain. They had no interdomain intelligence or cognitive fluidity. They were carving, cutting and preparing stone tools, but they never used that skill for carving a useful article or an idol which was symbolic of their group. They had no symbolic identification for their group such as a flag or some such symbol. As if their social, creative progress had stopped for all practical purposes, at a level as it was 250 to 300 thousand years back, when their species was evolved.

In fact given our present knowledge of how brain works, we have much ignorance than knowledge about it. Assessment of linguistic or cognitive abilities of another species must be based on their lifestyle, hunting methods, ornaments they ware and so on. We must rely on the archeological record to get some idea of what they could do. Whether use of fire was known to them, and how and for what purposes they were using it, and their burial rituals? From this we will know whether their brain size and functional neural structures were competent for regulating their manual, cognitive and linguistic abilities. We can make some general, though not detail inferences concerning language and thought process from the archeological records.

There was not much improvement in their hunting tools. Their predecessors, *Homo erectus* were using the tools found in the gorges of Oldovan river, Neanderthals could do some improvements in these and they were using tools which were known as livolise tools which were also used by *Homo erectus*. Only addition was they knew how to sharpen these tools and prepare a hand axe by fixing a wooden handle to it and by fixing a long bamboo handle to it and make a spear. However, they did not know how to prepare spears, which could be thrown from a distance or a bow and arrow. They had to approach the prey from close distance and there are evidences of their getting themselves injured because of attack from animals they were hunting. Tools and tool making technologies have got much more importance in deciding the progressive state of the species. Ian Davidson is an Australian archeologist who has an encyclopedic knowledge of the human tool record, has argued that modern human language and thought appeared only 35000 years ago in Australia and therefore that is the period of migration of *Homo sapiens* to Australia. This he thought because of tools they were using. Which is revised later by further evidence to 60,000 to 40,000 years ago.

The methods of hunting of Neanderthals were also the same or not changed. They were going long distances for hunting and bringing the carcasses back to the cave. These efforts were time consuming and tedious. On the other hand *Homo sapiens* were changing their strategies according to season. The flocks of gazzelle deer were migrating from one place to another according to change in season. *Homo sapiens* were changing their camping sites according to season and were following the gazzelle deer. That saved lot of traveling time and efforts and helped getting more prey.

How these changing strategies were proved or was it just a guess work? No, this was proved by finding the teeth in the fossils jaws of the animals killed, and brought back. Just as there are annual aging rings on the bark of a tree there are seasonal enamel layers on the teeth. They change their appearance according to season. The enamel layers laid down in summer are a little more transparent. This can be identified when examined under the microscope. In and around the Neanderthal caves teeth of the animals found had lines laid down in all the seasons. While those found around the sites of *Homo sapiens* were all of one type as *Homo sapiens* travelled after them and all were killed in one season only. *Homo sapiens* were basically nomadic and followed the flocks of deer wherever they migrated. This is indicative of the difference in cognitive development of the two. The *Homo sapiens* were nomadic while the Neanderthals were cave dwellers. There were no built up fire places found near the caves of Neanderthals, they made use of fire but that must be, for protecting from cold and to frighten the predators, while near the camping sites of *Homo sapiens* constructed fire places were found and evidences are there of construction of some sort of huts to protect themselves from changing seasons.

During the period of Neanderthals the ice age of Europe was coming to an end. During such changing climatic conditions usually there are heavy storms and heavy rains that might have caused severe damage. If Neanderthal had language, there would have been verbal communication amongst each other in the group, and they would have found some solution for nature's unknown fury. They would have prayed some unknown spirit, might have prepared its imaginary idol and so on. But Neanderthal had no interdomain cognitivity required for such a thought and no such idol or any other symbolic article was found around there caves. *Homo sapiens* on the other hand, had an interdomain cognitivity and some such symbolic articles are found near their camping sites, such as rain God or wind God. Even today we can see such idols on the outskirts of tribal villages to prevent evil spirit from entering in.

A three cm long stone article is found near a Neanderthal dwelling cave. This must be a stone from volcanic eruption. Some people have claimed, they could see lines of a sketch of a lady. This was known as Berekhat Ram figurine. This is no more than 3 cm in size, found in a 250 thousand-year-old deposit at a site in Israel. Some claim the stone was deliberately modified into a female form with a head, bosom and arm. Some think any similarity between the stone and a female form is entirely coincident. It is more imagined than real, just the way we convince a child that there is a deer on this full moon day!

Another contentious artifact is incised fragment of bone from Bilzingsleben, the marks on which, Dietrich Mania, believes were a part of symbolic code. A few lines however do not make a symbolic code. An alternative and more likely interpretation of the Bilzingsleben lines is that they derived from using the bone as a support or board while cutting meat or grass. Those who cannot bear the idea that the Neanderthal lacked the symbolic thought, claim that they lived such a long time ago that few, if any, symbolic artifacts would have survived. Time is not however the only determinant of preservation. There are numerous extremely well preserved Neanderthal sites, which have provided many thousands of artifacts and bones, in addition to well preserved burials. And yet all that can be found at such sites, are a few pieces of scratched bones and stones.

As against that Christopher Hinshelwood of South African museum initiated excavation in 1991. Within a few years he had shown, Blombo's Caves to be the most important currently known archeological site for understanding the origin of modern human thought and behavior and by implication language. He found two striking specimens, small rectangular pieces both just a few cm long what he described a shell like ochre; and marked with cross-hatched designs (Fig. 5.6).

The patterns are quite different from Neanderthal sites such as Bilzingsleben. They are sufficiently ordered to rule out any risk that they were hatched by chance. Moreover as same design is repeated on two separate artifacts, the impression is of a symbolic code. They are dated 70,000 years BP Blombos cave articles are the earliest unequivocal symbols we have from the archeological records not only from Africa but also whole of the world from *Homo sapiens* period. Later however even older specimens were found.

A more challenging argument for the presence of symbolism comes from the possibility of Neanderthal body painting. Stone nodules containing mineral manganese dioxide have been found at several Neanderthal sites, powdered manganese dioxide mixed with blood, tree sap or water makes a black color. As the Neanderthals have left no traces of pigment on cave walls or artifacts, the most likely explanation is, it was used for body painting. Neanderthals having evolved during ice age were white-skinned, and they were big game hunters, which makes it entirely plausible that the paint was used simply to camouflage their bodies. Alternatively for sexual reasons to attract the mate. There is no evidence that it had any symbolic or cultural importance. Even though, red ochre color was available during that time they never used it.

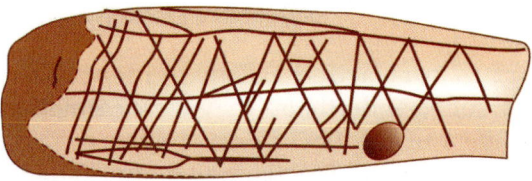

Fig. 5.6: Slab of ochre with symbolic design

On the contrary *Homo sapiens* used red ochre color extensively. The earliest representational art from Africa consists of painted ochre slabs from Apollo 11 cave in Namibia. These bear images of both real and imaginary animals. They are dated 60,000 years ago. Hinshelwood has found shell beads in Blombo's caves which are likely to have been attributed with symbolic meanings when worn 70,000 BP. Moreover there is compelling evidence that by 100,000 BP if not before, *Homo sapiens* were making and using symbols. Archeologists assume that the earliest use of 'ochre' was for body decoration—prehistoric cosmetic. This is not only because it is found in large quantity near their habitat but because of its red color. Red has been shown of special significance to human because of the range of the physiological effects it induces including change in heart rate and brain activity only by visual impact. The significance of red was explained by psychologist Nicholas Humphrey in 1976 in his brilliant essay "the color currency of nature." This is perhaps the reason why red is used for stop signal or danger signal on road signs. The red bluish on the cheeks of the girl friend invokes similar physiological reactions in the partner.

Ivan Turk and his colleagues have found a piece of bone in Dirge babe cave, it was about 114 mm long. He believed that it was flute and belonged to the period of Neanderthals. But such more pieces are found and holes on these hollow pieces are made by canines of carnivores. If Neanderthal knew about flute they would have certainly prepared more musical instruments.

As against in the period of *Homo sapiens* we have already seen an example of a flute found in Geissenklösterle dating 36,000 BP. As *Homo sapiens* had developed cognitive fluidity they not only made use of that as a musical instrument but also devised some more rhythmic instruments like drums, etc.

One fact is forwarded as an evidence of cognitive development of Neanderthals. They were caring for wounded or old members of the flock. Was it done deliberately? The newborn infants of Neanderthals were not independent just like *Homo sapiens*. This was because of their bipedalism.

Therefore they had to keep their newborn and their mothers back in the cave whenever they went out for hunting and bring back food for them and care for them. Automatically they must be caring for the old and wounded left back in the cave. Their single domain social intelligence was developed to that extent but there was no cognitive fluidity as no other signs of cultural development were shown.

Neanderthals became extinct about 30,000 BP. In the latter period of their existence some changes are seen in their lifestyle, they were using pendants, ear rings and chains of beads prepared out of teeth and bones of animals. Some such articles are found in their graves. In their latter period of existence they were seen deliberately burying their dead. The use of these ornaments might be because they had snatched them from their rivals *Homo sapiens* who had already crossed the border of Africa and entered Eurasia, or it might be left over from them or prepared by imitation. Otherwise if they had that cognitive ability why there should have been a dark period of 50,000 thousand years between their coexistence.

There are signs of migration of *Homo sapiens* from Africa to Eurasia round about 100,000 BP. At the graves found in Israel at Skhul and Qafzeh some symbolic articles are found. Head of a deer held by horns, by a dead, a flag, staff of a priest, and many others. These are certainly indicative of their cognitive development. They had an idea about unknown spirits, and life after death. Such symbolic articles are also found by Berty MacBury in Kubi fora and Blombo's caves. They are dated around 80,000 years BP.

Neanderthals became extinct from the world by 30,000 years ago. They had an evolutionary dead end. Why it was so? There is much of guess work about it. They had no mating chance with the new species as they had no language. This is found even today. Two groups who have got entirely different dialects do not mate with each other easily. The other reason is they could not stand the strategies of the new species of *Homo sapiens* who were cognitively more advanced, or they might not be competent to stand the changing climate of Europe. Whatever might be the reason for their extinction it is still a guess work.

Conclusion

All the evidence we have collected so far is all indicative of one fact that the Neanderthals were not competent to produce phonetic compositional language equivalent to *Homo sapiens*. They might be capable of producing some phonemes but were unable to produce important vowels [i], [u] and [a] and consonants like [k] and [g]. There speech also, if there was any, was nasal. All these factors rendered it unintelligible.

Their brain and neural development was not competent for compositional language. We have seen that their IPL was not fully developed as in *Homo sapiens*. IPL is the most important part of the brain in the origin of speech in modern human. The neural circuits competent for the utterance of all the vowels were not developed as they were not capable of uttering them. The nature is very conservative. The action potentials in any activity and the neural circuits competent for that activity always develop in tandem with each other.

The evidences of their cultural and social development are also indicative that they had only single domain cognition. Cognitive fluidity is essential for the development of compositional language.

All this circumstantial evidence points only to one conclusion that the Neanderthals were not competent of producing phonetic compositional language like modern human. As Neanderthal is the species which had evolved just before us there is no question therefore any other pre-human species having speech and language like that of modern human, and same evolved only with *Homo sapiens* about 150,000 years or so BP.

Homo sapiens migrated all over the world. Did they carry their language with them? If so why there should be about 6000 or so languages all over the world? These are the questions we will have to answer. Can it be that *Homo sapiens* evolved from the *Homo erectus* who had migrated before them at various places, which is called multiregional migration theory. We will answer all these questions, whether the modern human language developed in Africa only and why there are so many entirely different languages in the world in next chapter.

Out of Africa Migration *vs* Multiregional Migration and Why there are Many Languages?

It is now accepted by the anthropologists that Homo sapiens evolved in Africa and from there spread all over the globe. This is known as monogenesis. If we accept this view then there should have been only one language in the world. There is another view supported by few. Some pre-human species like Homo erectus had migrated to many parts of the world. Wherever they reached Homo sapiens evolved, from them at multiple places. This is known as multiregional theory. Therefore, there are many languages, each group developing their own language. Which theory is correct and where exactly the modern human uttered his first word. We will try to find out the truth in this chapter.

Peep in any household and you will listen this dialogue.
"Dear, you locked the door yesterday night, where are the keys?" The whole house then starts searching for the keys. Or
"Dear have you seen my mobile? It was in my purse when I came back from kitty party". "Why not look on the kitchen platform, or under the pillow in the bedroom?"

But our puzzle is not that simple to solve. Where did *Homo sapiens* utter his first word? As believed by most of the archeologists and anthropologists, sometime around 150,000 years ago when *Homo sapiens* evolved in Africa, the first word was spoken.

Albeit another opinion exists that wherever *Homo sapiens* evolved from *Homo erectus* who had migrated out of Africa 1.6 million years before to all the corners of the globe, which were explored till that period, the first word was spoken. That can solve our another question , why there are many languages in the world. As evolutionist Carl Zim says 'no one knows the exact chronology of the evolution of language.' Language leaves no traces on the human skeletons.

MONOGENESIS VERSUS MULTIREGIONAL MIGRATION

Linguistic monogenesis (the Mother tongue theory) is the hypothesis that proposes, there was one single proto language from which all other languages spoken today have descended. The multiregional hypothesis would entail that modern languages evolved independently on all the continents, a proposition considered implausible by proponents of monogenesis.

Homo sapiens evolved in Africa from *Homo heidelbergensis* and migrated out of Africa is now a well-accepted hypothesis, after extensive studies of fossils, by archeologists and anthropologists. Language does not leave back any material evidence. *Homo erectus* were biped and had settlements in Africa. One such settlement was identified by Glenn Isaac in the basin of Kubi fora river in south west Africa. They marked it as FxJj50. The settlement was near a stream. They found there fossil remnants and found about 20,000 pieces of bones of different species of animals hunted by them. Whether they were good hunters or only scavengers living on carcasses killed by other animals is doubtful. As the population increased, there was demographic pressure to migrate. As the group population increased they must have faced another difficulty, that is of communication. They were communicating with holistic vocal calls, but those calls were now insufficient to convey the necessary information, with each other. Alison Wray has proposed a theory giving a solution to this problem. She has suggested that they might have developed the capacity to segment the calls into small segments and combining them by the principle of permutation and combination, they might have increased their repertoire. This concept they might have picked up from birds. The birds break up their song in pieces and use separate pieces, or combine them in different sequences and increase their repertoire.

Why Migration?

Homo erectus had migrated all over the world. *Homo erectus* was the first pre-human species to migrate out of Africa. Fossils of *Homo erectus* have been found all over Eurasia, even to the East, as far away as China. Fossils are also found in Narmada basin in India and in north Western region of Pakistan. One group migrated along the coast of Indian subcontinent to islands of Indonesia (Fig. 6.1).

Why *Homo erectus* migrated? Demographic pressure was the main reason. One nonagricultural family needs about 250 hectares of land for their survival. If the group number increases more than 100 to 150 they need about 15 to 16 square km area for their survival. As there was no agriculture, the only source of food was fruits and other products of trees and hunting. They were called hunter-gatherers. And if the tribe increases still in number they have to migrate long distance searching for a new habitat. The other reason was changing climatic conditions and they were always in search of better climate which suited them. The third reason is inquisitiveness which took them places. The very thing that *Homo erectus* migrated is indicative that they were more advanced than the previous pre-human species *Homo habilis*. Their brain size was enlarged

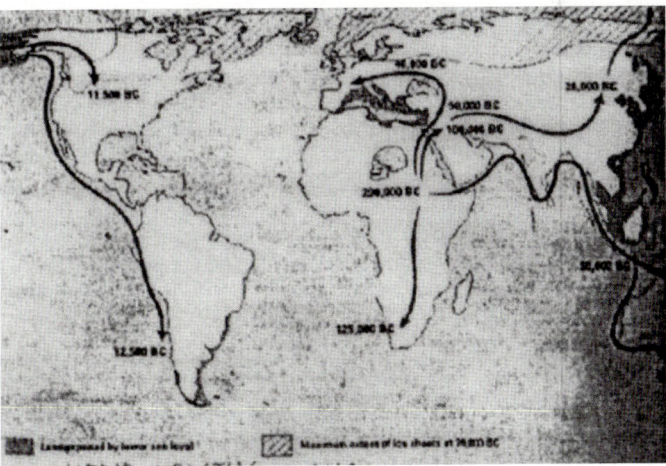

Fig. 6.1: Migration of *Homo erectus*

to about 750 to 1250 cc. They were biped and they had heel walk. That is why walking became easier and they could walk long distances. Their brain size was increased, is a sign that they have more neural circuits developed for new functions. They had better tools, they were using better tools than 'oldovan' tools. These were called as livollise tools. They were using hand axes. Still there are no evidences that they had interdomain cognitive development. There were no signs of any symbolic articles around their habitat.

Millford Walpoff of Michigan University and his colleagues have proposed that *Homo erectus* is a species which was evolved before *Homo sapiens*. *Homo erectus* had migrated all over the globe. Therefore *Homo sapiens* also must have evolved from erectus wherever they went. When anatomical and neurological structures of *Homo sapiens* developed to the stage of competence for evolving a language, they had their own language. That is the reason why there are many languages in the world. This is known as multiregional migration theory. This is perhaps very convincing explanation of the fact, why there are about 6000 different languages in the world.

There is one evidence that goes against this multiregional theory. When *Homo erectus* migrated they had reached islands of Indonesia. They had settled on one island called Flores, the species was named as *Homo floresiensis*. They were there up to about 12000 years ago before they became extinct. They never had any further evolutionary development. They reached an evolutionary dead end. It is most likely therefore that *Homo erectus* wherever they had migrated had reached an evolutionary dead end, and no new species evolved from them there. Perhaps this might be because of genetic isolation which is also considered as one of the causes for the extinction of Neanderthals.

Why Many Languages?

Mythology

No one knows exactly how many languages are there in the world partly because of the difficulty in distinguishing between a language and a dialect? One authoritative source that has collected data all over the world, 'The Ethnologic' listed the total number of languages as 6809.

Religions and ethnic mythologies often provide explanation for the origin and development of the language. Most mythologies do not credit evolution with the origin of language but take it as a language of God. The bible's explanation of the origin of multiple human languages is provided in the 'Tower of Babel' incidence recorded in Genesis 11:1–9. Scripture simply and confidently aserts: 'Now the whole earth has one language and one speech'. As the population increased, it apparently remained localized in single geographical region. Consequently no linguistic variation ensued. They were busy constructing a Tower on the outskirts of the city of Babylon. Their intention was to reach heaven the Abode of God. They did not listen to the command of God to disperse all over the globe. God punished them by confusing their tongues. No one could understand each other. There was a quarrel in different linguistic groups, the construction was abandoned and they dispersed all over the globe. The Babel account suggests that several languages came into existence that day.

Monogenesis

If it is agreed that *Homo sapiens* were evolved at various places wherever *Homo erectus* migrated, the evolution of *Homo sapiens* by natural selection would have reflected in differences in their physiological systems as a response to entirely different climatic, environmental conditions, food availability and fauna and flora around. However, no such difference is encountered. Anatomically and physiologically they have got the same systems working in the same fashion world over. Their auditory system is exactly similar and their SVT and articulatory system is the same. We will see latter that they respond also in the same way to the stimuli to these systems.

Fossils found near the delta of Klassi river at Hearty were studied for their DNA and when comparative study was undertaken with the *Homo sapiens* from all places around the world there was genetic similarity with those found in fossils in Africa. This is a proof that all the *Homo sapiens* were evolved in Africa only. Max Ingman from university of Upsala studied mitochondrial DNA from 53 individuals from all over the world and proved that all of them had their maternal ancestor line coming from Africa. Every cell has got mitochondria. In the sperm the mitochondria are in their tails. When fertilization of male and female gametes takes place the tail of the sperm is shed out therefore all the mitochondrial DNA in the cell come only from ovum and from the mother side. Naturally maternal ancestry can be traced from mitochondrial DNA. From this study he could show that all the *Homo sapiens* evolved in Africa about 170,000 BP (give or take 50,000 years). He thought, as language is a biologically endowed faculty to modern human the language gene FOXP2 also must have appeared at the same time by mutation. Obviously FOXP2 was also found in human wherever they spread. It is not however accepted as the language gene.

In recent years it would appear that the 'out of Africa model' is more acceptable to researchers in this field. The genetic studies of Cavalli-Sforza and colleagues place the basic split in the human tree between sub-Saharan Africans and everyone else, and they date this split at some time after 100,000 BP. Studies of mitochondrial DNA have posited the same basic split. Recently a team of geneticist led by late Allan Wilson had been investigating human phylogeny on the basis of mitochondrial DNA which is inherited strictly, along the female line and therefore easier to trace geneologically than nuclear genes. They found as did the Cavalli-Sforza team using nuclear material, proteins and enzymes for their study, that the most basic cleavage in the human population separates sub-Saharan populations from the rest of the world.

Universal Culture and Language Base

If human species evolved at many places there would have been basic difference in their culture and language. But it is not so. Inspired by Chomsky's universal grammar (UG) Brown has tried to characterize the universal people (UP). He has scrutinized archives of ethnography for universal patterns underlying the behavior of all documented human cultures and languages. Brown was able to characterize the universal people in gloriously rich detail. His findings contain something to startle almost anyone. The basic characters of all the languages are the same. All the languages have phoneme as the smallest unit of the language. The number of phonemes is always limited. We can have unlimited vocabulary from those limited number of phonemes. For each phoneme there has to be neural circuit in the brain. This is very expensive and nature is very conservative. It cannot afford to have too many phonemes. The orthographic punctuations and the prosodic factors are almost the same in all languages. The ability to differentiate between voiced and unvoiced phonemes is present as an innate gift for infants of parents belonging to any language culture. This differentiation depends on the difference in msec in the utterance of a phoneme and the onset time of vocal cord vibrations. For example [p] and [b] are both labial sounds but [p] is unvoiced and [b] is voiced (more about this in Chapter 10). This ability is present up to 10 months of infants age for any language not depending on environment of the language of the parents of the child. Werker and her colleagues did observe this in case of children of 6 months old, born of English speaking parents. They were given to listen pairs of phonemes in Hindi and English language, pulse rate and sucking rates were used as paradigms for comparison. The infants had not heard Hindi as it was not in their environment and English as that was not their age to learn language. Even then they could recognize the difference. As this ability is innate and therefore biologically present. It is a proof of the fact that all the *Homo sapiens* were evolved from one species that is Heidelbergensis in Africa. This is a proof that all the languages and cultures in the world had beginning at one place and that is, in Africa. According to Brown the universal people and language have the following common features. There are words, for days, months, years, past, present, future, body parts, inner states (emotions, sensations, thoughts) behavioral propensities, flora, fauna, weather, tools, space, motion, spatial dimensions, physical properties, giving, lending, affecting things and people, numbers (at the very least one, two and many more), proper names. Distinction between mother and father, son, daughter and age sequence. Binary distinctions between male and female, black and white, natural and cultural, good and bad, and so on. The unavoidable inference is that sense of similarity must be innate. This much is not controversial. The only conclusion we can draw from all this discussion is the basic language was originated or rather its roots are somewhere in Africa only, along with the evolution of *Homo sapiens*.

Is Indo-European Language the Oldest Language?

In 19th century European linguists proposed a thought that Indo-European language is the oldest in the world. All other languages are on a primary stage. They also proposed that some of the languages like Khosian language in Africa, are still somewhere, in between, the vocal calls of chimpanzees and human language. They still use clicks as a part of language. English use clicks as interjections. In English, for example dental click [/k] is written as 'tsk, tsk' and expresses mild disapproval. The lateral click [//k] is used by English speaker to urge on horses, or to call cats and dogs, thus they are part of prosody. In Khosian languages, however, clicks are used not in this imitative fashion, but as ordinary consonants that combine with other sounds to form words. In 20th century however linguists came to the conclusion that no community in the world is in regards of cognitive, and linguistic progress is backwards.

Here I cannot resist myself quoting a story.

In our schooldays we use to go on a picnic in winter days in jungle near our village. On the top of the mountain there was an old temple where we use to stay overnight and enjoy cooking and games. Once on the way we saw a cluster of bananas hanging from a banana tree. The ripped banana cluster was very attractive and we just thought, it is a gift from 'Jungle God' to us. We cut and took it with us for night feast. By evening though, some tribes came and told us we have stolen their banana cluster. We tried to convince them that we have brought it from the town below. And they did an amazing thing, they had brought with them the part of the stem from where we had cut it and they showed us how it fits to the cluster stem exactly. We were speechless and stunned and quietly handed it back to them. How can we say that their cognitive progress was anywhere less than us? Their language was also developed enough for their environment.

They could think of past and future. They could think of fruits they saw yesterday will be ripened in future. They could think of things which are not in presence. They could think of the things in other world and powers unknown. Rains are good but rains can be torrential and destroy, so they thought of a Rain God, "Pavasha" and worshiped the God. Their language vocabulary might be limited but that does not mean that they were not capable of developing their language as the need be. I remember of a smart tribal boy in way back in 40s. We tried to teach him and kept with our children. In a few months he started talking in our language. This is an evidence that they have the inborn capability of learning any language. The Maori tribe in New Zealand have got only 8 consonants and 5 vowels in their language. They could organize about 91, 125 words out of those phonemes. William who stayed with them for one year to study their language found out, they were using only 12,000 words. Those many were enough for their requirements. When in 8th century they came in contact with English people their language also started developing fast.

The above discussion proves beyond any doubt that all the *Homo sapiens* evolved in Africa and when they migrated out they had language, though not as full-fledged as it is today. Still we will have to answer one question and that is if *Homo sapiens* migrated out of Africa with a language which was developed to some stage and why there are so many entirely divergent languages in the world?

Why Many Languages?

We have already seen that the Cavalli-Sforza and his group place the basic split in the human tree between the sub-Saharan and everyone else. Furthermore the greatest lingusistic diversity appears in Africa. There were four major families and each had separate habitat, each group of individuals had their own tongue. When Americans entered Africa with their more advanced weaponry, to get slaves they took away these tribes with them as slaves to work on their farms. Americans perhaps knew the story of tower of Babel. Shrewdly they divided the individuals speaking different tongues and employed them in different farms. No one knew each other's tongue. They use to communicate with each other via a language having a very, limited vocabulary and the rest was conveyed by gestures. This was known as 'Pidgin' language. Pidgin was significantly simplified language with only rudimentary grammar and a restricted vocabulary. In their early stages, Pidgins mainly consisted of nouns, a few verbs and adjectives. It often happens with us also. If we go to another country and do not know the local tongue, we have to manage with a few words and the rest with gestures. This is as good as Pidgin. When tribal people come from their remote villages to the town, they manage to communicate with the people in the town in the same way, though they have got their own language which is matured enough for their needs. What happened to Pidgin languages? The next generation of these slave communities was not knowing even their ancestral tongue. They picked up some words from the Pidgin and some words and grammar from the farm owners with whom they came in contact. Thus, they developed their own language. This is known as 'Creole' language' Pidgin is a simplified language that develops as a means of communication between two or more groups who do not share a common language. Did all children grow up speaking a Creole language? The answer is they did their best to do just that. People around them however persisted in English, French or some other language and the child had to modify the grammar of the native Creole until it conform to that of the local tongue. Is it possible that some such thing might have happened to *Homo sapiens* (Fig. 6.2)?

The modern man had competence to learn language. Their anatomical structure and neural development had already progressed in that direction. They had competence to produce phonemes and the words related to a particular object or action. The words were accepted by the individuals in the group and that was developed at least as a Pidgin language. Which

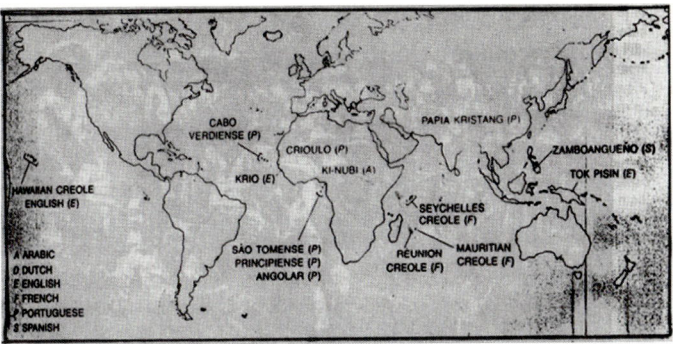

Fig. 6.2: Pidgin of *Homo sapiens*

gradually must have progressed to a language which was similar to Creole. This latter developed, into a matured language of today. This process was not completed in just one generation but must have taken a few thousands of years and many generations. Even then the question remains if the language matured that way from one single language why there are so many entirely different languages in the world?

One reason is after sub-Saharan split *Homo sapiens* migrated below Sahara, and out of Africa, not in one major group. As there was a demographic pressure and groups became larger since 150,000 BP they migrated in different groups from each habitat separately out of Africa. From about 250,000 BP to 50,000 BP very few fossil remnants are available and it is difficult to trace the detail history of their movements during that period. We have already seen about the fossil remains found at Skhul and Qafzeh in Israel and their cultural advances. Similar remnants are found at Kubi fora and Blombo caves dated about 70,000 to 80,000 BP. Which indicates if we accept that *Homo sapiens* evolved by 150,000 to 200,000 BP, they migrated at different periods in separate groups, with whatever Pidgin or Creole language and cultural development they had?

There is one more possibility. Groups which migrated for long distances were isolated. They might be having some Creole language already developed and further development was depending on the environment, needs, flora and fauna and the lifestyle they had in those locations. This is how they might have evolved their own language. If they happen to come across with some other earlier migrated group there might be mixing of the two languages and they had a new one. One such sites have been identified. The languages they had were not only competent for their lifestyle but they had made scientific and interdomain cognitive progress.

In early 20th century the explorers thought that they have explored all the human habitats which have developed their language to a state that they are having rich culture and even having scientific advancements and thinking of their environment. But had they explored the world thoroughly? Just look at the following story.

In 1930, Michel Lahi and his colleagues reached New Guiney island in south east in search of gold. This island is surrounded by Pacific ocean and in the center there is a high mountain. Some missionaries had already reached the island but they had not traveled beyond the coast line. When there was a news of gold find on the island, Lahi and his colleagues thought of going beyond the mountain. They went atop the mountain and rested there for the night. What they saw at night? There was a large basin formed by several small rivulets and there were clusters of lights seen at many places. This was a clear indication of human habitat. When in the morning they went ahead they could see small villages. They had to face aggression in the beginning by the locals on finding strangers. But when they develop friendship by lot of skill and gestures they found to their surprise the villagers had a language of their own and that was rich enough and sufficient for their lifestyle. There were about 800 settlements in that basin. This must be a migrated group from Africa long time back and remained isolated. They developed their own languages for communication.

Homo sapiens reached Australia sometime between 60,000 and 40,000 BP. Later English people reached sometime between 6th and 8th century. It was found by exploring deep inside the land, that there were tribal villages and they had their own tongues but they had lot of similarity and genetically also they were identical.

Even in India, we can easily observe that the languages in south India are entirely different than those in north. The south Indian languages are grouped as Dravidian and the African emigrants had reached there separately by sea route. The north Indians however reached via Khaiber pass and their language was Indo-European. According to shri Chintamanrao Vaidya, some Aryans entered the Ganga-Jamuna basin via north east route. Kauravas in Mahabharat were one of them. Languages Brahui and Tamil belong to Dravidian family which is found primarily in south India. Brahui however the most divergent language in the family was in Baluchistan province in the west part of Pakistan and appears to attest to an earlier, more widespread version of the Dravidian family, before it was overrun by speakers of various Indo-European languages. There are believers amongst linguists who believe that Tamil language is older than Indo-European and it has great influence on Sanskrit and Marathi.

This way the *Homo sapiens* from Africa spread all over the globe with their language developed to various stages from Pidgin to Creole and to still more advanced stage of language and settled at various places, which they found suitable for their survival. There they developed their language further depending upon the environment, flora and fauna, food, and mutual dialogue, and cultural advancement they had, after they found the site suitable for a long time settlement. As they migrated in different groups, and settled at various places, naturally the languages they developed were different from each other. Modern human species have diverged the meaning of the phonemes and words formed, from the actual utterance according to whatever they thought at that time and was accepted by the society. The articulatory system was however the same and the phonemes uttered were somewhat similar. The meanings of the word, the rules of syntax however differed as the languages at various places developed independently. That is the reason why there are so many languages in the world. The actual phonemes and their number may be different in each language and as the groups divided and migrated to many different regions of the world their languages also might have some similarities with each other. Thus, there might be a mother language from which many languages developed at various places. How many such language families can be identified and what is the route along which they spread with variations, is a different discipline of linguistics called typography and beyond the scope of the present treatise.

Conclusion

From all this discussion we can certainly reach one conclusion that all the *Homo sapiens* belong to one family which evolved in African mainland and had this miracle of speech and language basically evolved only in Africa. Their split between sub-Saharan and the rest of the world is now generally accepted. The languages there after diverged from each other so much that it became difficult to trace their families and roots. As mentioned above it has become a separate discipline in linguistics and diverse opinions and theories are forwarded by experts.

We have thus seen how the modern man speaks, when he achieved this art, and where he uttered the first word. And then even if the articulatory system man got as a result of evolutionary process is similar everywhere, who decides what word to be used for what object or concept? Formation of syntax, and concept of semantics certainly is not a gift of God. In this scientific era we will have to have answers to all these queries which are beyond any doubt (Cartesian doubt), by scientific reasoning. Let us see how we can succeed in that. However, we will have to go through the thoughts of many scientists, mainly linguist in the next chapter Also their ideas about pronunciation of first word before going through my theory of origin of language.

7 The Origin of Speech and Language: Some Theories

We have already seen how the man speaks. About the articulatory system producing the phonemes, about different parts of the brain where the speech is organized. Let us review again in short about the different parts of the brain and their evolution (refer Table 3.1 in Chapter 3).

Speech and Language

Language is not one package, it has got words, sentences or syntax and these sentences together are organized to give some information, all this is called semantics. Let us first see what are the special characters of the human language?

What is Language?

1. Our vocabulary is enormous. A child of 8 years old has vocabulary of about eight thousand words.
2. Only humans have got function words, that exist exclusively in the context of language. Words like dog, night, or cat are nouns which refer to actual things or events. We only can make use of conjunctions such (–and, if,–), to organize language with compound sentences. Conjunctions have got no independent existence.
3. Only human can use words "off line," that is to refer to things or events that are not currently visible or present, they exist only in the past, in future or a hypothetical reality. An example can be given "I saw an apple on the tree yesterday and decided, I will pluck it tomorrow, if it is ripe."
4. Only humans can use metaphor or analogy. We can make use of puns and construct poems, just as Tagore's

description of Tajmahal "Tear drop on the cheek of time" as quoted by Prof Ramachandran.

5. Flexible recursive (complex) syntax is found only in human language.

Ancient Perspective

Considering all these perspectives of language, linguists since ancient time such as Panini, Patanjali, Katyayan, till in recent past, like Saussure, Sapir, Bloomfield and others thought of descriptive linguistics or structural language only. We have already seen about the work of Panini and other ancient grammarians and their amazing work documented 400 BC in Chapter 3.

Recent Past

Swiss Linguist Ferdinand de Saussure (1915/59) divided language into two levels. A level of linguistic forms (La Lingua) and a level of phonetic substance, or in effect observable speech (La Parola). The important level for him was the level of linguistic forms, which are present a priori or from birth and actual speech or 'La Parola' of minor significance. La Parola was called also as structural linguistics. The main focus of structural linguists was phonology. They call this part of language which is uttered out of the mouth as 'La Parola'. He had interest only in La Parola. He was having the same opinion as that of Noam Chomsky, that La Lingua is innate and is present from birth. Some other contemporary linguists such as Sapeer, Bloomfield and others also gave importance to only structural linguistics.

Linguists gave a deep thought to the structural language, but no one thought of giving importance to the search of how language is composed in the brain and ordering its production or La Lingua. In 1859, Darwin had already proposed his theory of evolution by 'natural selection,' though its trustworthiness was hotly debated no one believed that language was also acquired by 'natural selection.' We have already seen some of the opinions about the origin of language proposed prior to, Darwin's theory of evolution was published, in our introductory chapter. Lavelt said that origin of language is the most neglected subject of human physiology.

Twentieth Century

The study of language from 19th century onwards has rather neatly followed the same course as Schopenhauer's aphorism. The 19th century philosopher Arthur Schopenhauer said "All truths pass through three stages. First, it is ridiculed. Second, it is violently opposed. Third, it is accepted as being self evident".

Linguists once considered pursuing the topic an absurd endeavour. Then it was banned. After that the official ban developed fairly seamlessly into a virtual ban. Now where once several researchers proclaimed that you cannot study, many say you can, including the scholars best known for saying you cannot or at least you should not bother. Its remarkable, now that the rhetoric about language evolution has shifted, how quickly, what was once heretical has received wisdom.

In twentieth century some more theories were proposed. Many more linguists were still skeptical about the fact, that language is, very complex consisting of words, syntax, meaningful dialogue, prosody and so on, and therefore is it possible for such a complex ability to be evolved by natural selection or, is it really a God given gift? They gave example of an eye. If there is no lens what is the use of retina. Nature only selects a variable which is supplementary to the one present earlier and beneficial for survival. Therefore eye must be also innate.

Evolutionary biologist Richard Dawkins, who was a staunch follower of Darwin gives an intelligent explanation for this. We see many stages of evolution of the eye even today, from unicellular organisms to insects, and still more evolved animals like reptiles and vertebrates. Millions of years ago a photosensitive spot might have evolved in some species on the skin of that organism. Even today there are species like, some species of leeches and starfish, who's entire skin is photosensitive. To protect this sensitive portion, it went into a depression, that depression was filled with a jelly-like substance to protect that sensitive part from sea water. A transparent sack filled with some transparent fluid covered the depression to protect the jelly from slipping out. This might be the future lens. Eye can develop this way through many complex steps all taking place in sequence, by natural selection, over a period of thousands of years. Our ear is an another such example.

Many linguists in recent past have put up many hypotheses about the evolution of language. Let us look some of them.

Theory of Alfred Russel Wallace

Wallace, was contemporary of Charles Darwin. Independent of Darwin he had thought of 'Theory of evolution' by 'natural selection' and was to publish his hypothesis. But before that Darwin published his famous book "Evolution of species by natural selection." Though Wallace's thoughts were matching with those of Darwin as far as evolution of living world was concerned, he did not agree with him about evolution of language. He thought such a complex miracle unique to human could not have, evolved by natural selection. He also believed that language was a God's gift to human.

Theory of Noam Chomsky

World renowned linguist Noam Chomsky, who is considered as father of modern linguistics, was influenced by the dominant classical view initiated by Plato and reinforced by Descartes, that knowledge exists in human mind a priori (before birth). Noam Chomsky subscribes to this assumption and bases his linguistic theories on it. He has also embraced the classical tradition by referring to the problem of language acquisition as 'Plato's problem. Plato had said that how it comes, that, human being who's contacts with the world are so brief, personal and limited, is nevertheless able to know as much as he knows about the world (Chomsky 1986). Chomsky has emphasized poverty of the stimulus, and the insufficiency of experience as a basis for the difficulty in origin of language by natural selection. therefore consequent necessity for the language to be built in, a priori. He believes our innate knowledge of the language includes both the syntactic aspect (sentence structure) of the language and phonological aspect (sound structure underlying speech). He says this is a specific human ability, independent of intelligence, unbounded in scope and stimulus free. It also is the basis, for the unique capacity of possession of generative grammar. Generative grammar ideally specifies a pairing of phonetic (sound level) and semantic (meaning level) representation of language on an infinite range. It thus, constitutes a hypothesis stating, how speaker and listener both can interpret utterances, unless they are present as priori. 'At the base of generative grammar of any language there is a universal grammar a set of rules and expressions that Chomsky believes are part of the genetic endowment of every human irrespective of his individual language'.

Because of Chomsky's theory of generative linguistics or grammar in the last half of the 20th century the orthodox discipline of the structural linguistics had taken a back seat. Saussure placed linguistic forms, at sociocultural level which Chomsky described as E-linguistics. And he himself placed linguistic forms within the mind, present from the birth, as I-linguistics.

Chomsky's signature claim is that all human share a "universal grammar" otherwise known as UG, that can generate a set of rules of the syntax of every human language. Those committed to Chomsky believed that universal grammar exists in some part of our brain in the form of 'language organ' that all humans have, but no other animal has. Thus, the way language was defined by generative linguistics or universal grammar made its evolution more incomprehensible.

Chomsky does not believe that universal grammar evolved for the communication at all. He believes that it originally functioned as what he called linguistic expressions (Chomsky 2000) which allowed the possessor, in some sense, to talk to himself. Chomsky does not believe in the power of adaptation to produce new functions. But from the present stand point, we must raise the question of what would have been the adaptive benefit of talking to ourselves via universal grammar. One problem is, without incorporating meaning into universal grammar which consists only of syntax (sentence level) and phonology (sound), what would have been talked about? In addition, what would have been the phonological component for, if there was no transmission of linguistic information to others? We speak to be heard in order to be understood (Jacobson et al 1969). Thus, for Chomsky linguistic component is a set of ideas underlying language, which are naturally innate like Cartesian mind. In contrast performance has to do with what actually happens when we speak and listen. We speak by articulatory system and hear by auditory system or ear. Both are products of natural selection of evolution. Therefore, the innate ideas of universal grammar and actual performance of production by bodily organs is like mind and body bifurcation in production of language. The production is by a system, namely articulatory system, which is a product of evolution by natural selection. Actual performance by organs which are products of natural selection and generative linguistics being innate, naturally makes generative linguistic incompatible for Neo-Darwinism. But Chomsky's concept here, is both production of speech and perception mechanism must make use of underlying system of generative rules which are innate. So this does not match with the Chomsky's idea of innate rules for both, articulatory system which is a Neo-Darwinian concept and human ear which has developed by natural selection, on one side and universal grammar on the other.

It was believed in the early days that the universal grammar or the 'language organ' was hardwired in the human brain. Any one born with UG was born with the potential to learn any language. The working of the language organ was also thought to be completely separate from other parts of the brain. They were separate from the context of the spoken language and they were also completely separate from the similar systems, like music. This model of language was consistent with the general theories at the time, about how the brain functioned, namely a series of separate boxes, each of which computes different parts and different functions.

Chomsky's critics claimed that he chose data to support his theories, but then discarded it when it no longer suited, and that he intentionally misinterpreted his adversaries and then launched an attack against his own misunderstanding. Pinker and Jakendoff charged Chomsky with having abandoned the last twenty five years of his research and coopted ideas from models once he had completely dismissed.

Certainly no one new whether language was the function more of physics than behavior or biology. Instead of resulting from adaptation of some other existing function and by natural selection, language may have arisen as a byproduct of a very complex mental machine.

On the other hand Chomsky said brain was messy, how can such a messy organ can develop something so perfect, was a mystery, he said, that was for the time being insoluble. For many years he deemed language evolution unworthy of investigating. Chomsky was further saying for decades, indeed it was the enormous complexity of language that made it hard to imagine, not merely how it evolved but that it had ever evolved at all. He spoke often of innateness. The implication was the 'language organ' was specified in the genome and it was generally assumed that there was a gene or genes specifically for language.

The problem comes when giving explanation how the innate rules, which is a genetic endowment of human, can be competent for the production and perception of different linguistic structures (syntactic and phonetic) for different languages. Here Chomsky favors the botanical growth metaphor of eighteenth century philosopher, James Harris. "It is the internal vigor and virtue of the tree that must ripen the juices of a fruit to its maturity, though external factors like environment, soil, and many others may co-operate. This is just like maturity of bodily organs like liver, heart and so on". Perhaps here what Chomsky was missing, more powerful sociocultural effect on the maturity of language and perception, which is not required for maturity of liver and heart (Chomsky 1966).

Here we must make a mention of a recent major discovery important enough to have already come in introductory text books (Gazzaniga 2003). The environmental input influences the expression of genes, it can cause genes to become active. And genes influence the organism to react with environment.

This proves that external factors are very important in production and perception of speech.

Chomsky developed his concept of universal grammar by studies on adult languages. But then why a child starts having speech after the age of one and half to two years of age, if it was there as an innate faculty. Was there any need of development of any additional structures for production and perception (by natural selection)? Chomsky's explanation to this is, there is no need of external structural changes, but he says infants speak late because by that age, they are matured to the point when their innate capabilities enable them to do so. And what is the evidence that they are matured? The evidence is that they are able to speak.

His another view is maturity is like clearing the mist from mountainous landscape. The mountains are there but it is the mist that obscures the view, as the mist starts clearing mountainous landscape gradually begins to develop, things are there they are only becoming manifest.

By the time Chomsky was putting forward his views about evolution of language. General acceptance of the theory of evolution by natural selection of Charles Darwin and Neo-Darwinism, had been granted acceptance on a larger scale in scientific world. This acceptance of the importance of history forced Chomsky (1988) to deal specifically with the evolution of speech, which his predecessors like Plato and Descartes were not required to do. He dealt with it, still by denying the relevance of Neo-Darwinian theory of language origin, and putting forward an instantaneous evolution of language, by arbitrary, physical laws, as yet undiscovered. He gave an ingenious story.

"It may be that in some remote period, a mutation took place that gave rise to an infinite property to biology of brain cells, for reasons that have to be explained in terms of properties of physical mechanism now unknown and applying to the brain a certain degree of complexity" (Chomsky 1970, language and problems of language). Perhaps what he wanted to say was, that a language organ or Language Acquisition Device (LAD) was placed in the brain, which is quite possible considering the billions of cells present in the brain.

Chomsky perhaps to humiliate evolutionist, told a fairy tale. "It is almost as if there was some higher primate, wandering around, a long time ago, and some random mutation took place. May be after some cosmic showers which reorganized the brain implanting a language organ in an otherwise primate brain." He further adds "This is not a story to be taken seriously but it might be closer to reality than, many other fairy tales that are told about evolutionary processes including language by "evolutionaries"" (Chomsky 2000, the Architecture of Language).

Of course he could not explain when a language organ was implanted, It was never found by anatomist or tested successfully by neurologist. Moreover many questions will have to be answered. Why the cosmic showers only showered rays for implanting a language organ? Why they chose only the left brain for implanting universal or generative linguistics?

We have discussed Chomsky's views at length and in detail as they were dominating last few decades of 20th century, and still there are linguists who are believers of his hypothesis. Chomsky is so dominant about his view that he says "To say that language is not innate, is to say that there is no difference between my grant child and a rock or a rabbit." Thus, Chomsky spoke often about innateness but when you involve innateness it is hard not to make a few assumptions about genetics and evolution. This really was the 'language organ' or 'universal linguistics, as it was specified, must be in the genome, one will have to assume that there was a gene or genes specifically for language. But then he also saw language as a formal system (mathematical). Again for mathematical entity must have appeared from nowhere, with no precursor. This really contributed to wide spread view that language evolution was impossible and it must be a miracle. Chomsky for many years deemed language evolution unworthy of investigation. Against the backdrop of Chomsky's lack of interest, the problem of language evolution remained during the most of the 20th century, domain of the occasional crack-pot and a few determined and brilliant mavericks.

Big Bang Versus Gradual Adaptive Evolution

Some theorists think human language is absolutely unique and tempt to argue that it must have emerged through a big bang (somewhat similar to—Big Bang—theory of origin of universe) and not through usual serial mechanisms and chance events. The specialization of *Homo sapiens* and origins of language are surely two sides of the same coin (Crow 2005). Crow searched for major genetically controlled distinction between humans and all other species which could validly be associated with the emergence of language. Crow sites considerable asymmetry of the human brain. Certain areas are certainly developed more on the left and certain on the right. This clearly shows lateralization of modern human language, demonstrable with modern scanning techniques. This goes in favor of genetically controlled evolution.

It seems that the present archeological climate is shifting away from a revolutionary perspective towards ideas of gradual cultural innovation. And gradual adaptationist view of evolution of language. The scientist investigating this, think there is likelihood of having hundreds of genes which through mutation and potentiality of selection in the human lineage having contributed in their small way to the gradual development of the human capacity for the language. Some of the structures of these might improve the production or perception. Some may make the brain bigger and allow more skills to develop. Either way to a gradual approach is more in keeping with the usual forces of evolution, which tend to operate by changing and building on, what is already available rather than starting from nothing and trying for a quick fix answer?

Pinker diagnoses accurately the tendency of some theorists to see human language as absolutely unique, but also consequent temptation to argue that it must have emerged through a virtually unique circumstances, such as a big bang. We wish language our critically unique property to be, so amazing that it must have evolved through a big bang.

This is exactly the approach taken by Hauser, Chomsky and Fitch (2005, 2010) who identify a particular feature, namely

'Recursion' (recursion means a complex sentence or one having multidomain components). The only unique human component of the faculty of the language which they designated as FLN or faculty of language in narrow sense. Opposed to that was faculty of language in broader sense (FLB) which included all other faculties which were not specifically evolved for language but were essential such as respiration, vocal imitation, sound production, etc. What else is in FLB? The authors are vague on this aspect as to what forms the part of FLB and what does not? They are all brought together, just as in construction of an arch, by a 'key stone' and that key stone is "recursion" and everything that remains as a part of the arch is FLB.

By this time theory of evolution by natural selection proposed by Darwin was receiving acceptance on a large scale by researchers. Chomsky and others therefore stated in so far as there is any contribution of natural selection to the evolution of FLN it may reflect an initial development of recursion for nonlinguistic reasons, to solve other computational problems such as navigation, social relationship, etc.

The question then arises as to how and why this computational technique was extended, to language in human and then became uniquely human. Moreover, it is possible to have a recursion free modern human language as that of "Piraha" a tribal population. Their language has got no recursive sentences. For more discussion of this subject, refuting the role of recursion as role of 'key stone' and as being uniquely human element for language please refer to Jacendoff and Pinker (2005) "cognition". This approach of recursion as uniquely human is strongly colored by Chomsky's theoretical commitment of minimalist (one only-key element) program in which syntax and notably recursion is emphasized and other aspects of language are played down. Thus, it seems fair to say that 'recursion only' approach of Chomsky and others is in trouble. The FLN seems to include something more than recursion and perhaps no recursion at all. What then might have motivated them to propose a hypothesis in the first place? Pinker says it emerges from a wish to dissociate from adaptationists accounts involving natural selection.

Theory of Stephen Jay Gould

The next theory was proposed by one of the most distinguished exponents of evolutionary theory, Stephen Gould. Stephen J Gould based at Harvard was of the opinion that scientist were too quick to apply evolutionary explanations to everything. Since external features of our body did not result from adaptation, he argued, but are just accidental byproducts of some other evolutionary changes. Gould called these as biological artifacts or "spandrel". What is a spandrel? Since organisms are complex and highly integrated entities, any adaptive change must automatically "throw off" a series of structural byproducts just as an architectural spandrel, a triangular space between the top of the two arches on a wall meeting the ceiling, resulting as a byproduct of the design of arches (Fig. 7.1). They may be later coopted for some useful purpose like carving a design, but they did not originally arise as adaptations for some definitive use. When someone asked Gould, about evolution of language, he was uninterested even a little annoyed by the question. He waved his hands about and said "It is probably a spandrel. He argues that language did not evolve with a specialized purpose for communication. It is not either a specialized brain module, on the contrary it is a part of a more general mechanism that evolved earlier for other reasons, namely thinking. Only later did this system get repurposed or extended into a means of communication? In this view then, language was an exaptation—a mechanism that originally evolved for one function and then provided the opportunity for something very different (language), to evolve.

Fig. 7.1: Spandrel

In this view then thinking is an exaptation (borrowing from something which was present earlier), but the exaptation itself must have evolved by natural selection (from what)? Failure to appreciate this has resulted in much confusion and bitter academic disputes. The real example of an exaptation is the evolution of feathers from the scales of reptiles for insulation but as the birds were warm blooded they did not need that much insulation and the feathers were used for flight. Fingers were originally evolved in apes for having a good grip for climbing trees but modern man adopted (exapted the motor action or use) them for playing piano, pointing, counting and many other such uses.

Gould argues that thinking evolved first for its usefulness in dealing with the world. Later language was rooted in the system that gave our ancestors a more sophisticated way to mentally represent the world. Here again if language was useful for mentally representing the world, which was understood by thinking first, for what the thought was to be mentally represented? Obviously for communicating to others, or was language to communicate the thought to mind only, what was that language like? Was it just a byproduct? It was certainly not the language as we understand it today. Today's language consists of words, sentences and semantics. Such language was not certainly introduced in the system before appearance of *Homo sapiens*. Even animals can think, then what language they must be using to communicate their thoughts to the mind. Perhaps what Gould meant by 'language' was a means of communication of thought to the conscious mind. In fact we know very little about thinking and how it was evolved than we know about language from his description of it? Therefore saying language evolved from thought does not tell us much. As Dr Ramachandran says you cannot get very far in science by trying to explain one mystery with another mystery.

Theory of Steven Pinker and Bloom

Distinguished Harvard university linguist and disciple of Noam Chomsky declared language is an instinct just as coughing, sneezing, but not that simple. It is highly specialized brain mechanism, an adaptation that is unique to *Homo sapiens*. It is evolved by natural selection for communication. It is not an exaptation as Gould says. He agrees with Chomsky that language originates from a highly specialized 'language organ'. It evolved by natural selection is correct but it is incomplete. We do not know, from what this natural selection got started and then evolved to its present level of sophistication. In considering the evolution of any complex biological system (whether ear or the "language organ") we will like to know, not merely that it was done by natural selection, but exactly how it got started and then evolved to its present level of sophistication. Pinker and bloom wrote, Chomsky the world's known linguist and Gould the world's best known evolutionary theorist have repeatedly suggested, language may not be a product of natural selection, but a side effect of other evolutionary forces such as increase in overall brain size, and constraints of as yet unknown laws of structure and growth, or a spandrel. Overwhelming impact of research paper of Pinker and Bloom was, as if a door had been flung open. From that point on, more and more researchers felt that studying the origin of language was a legitimate academic inquiry after a hundred years or so of uncomfortable silence it had become intelligent to wonder aloud, how on earth we had come to be a species with word?

Charles Darwin

His views are 'It is my intention to give an account of the evolution of speech that unflinchingly adheres to a perspective—that contends in short, that speech did not just "happened" by means of a secular miracle but, instead evolved by descent with modification in accordance with the principle of natural selection.'

P Lieberman

Lieberman himself was a student of electrical engineering. However, he became bored with transistors and circuit boards and decided to take a linguistics class. He found the class exciting, for he liked language and was excited by the idea of using it to understand the mind. His path soon changed by observing the first speech synthesizer and the engineer in him soon got interested in knowing how speech actually works? He moved away from thinking of the abstract properties of the language and also from thinking of Chomsky, who's students initially he was. Once he started investigating the biophysics of speech he realized even big apes who are so near to human species cannot speak. He discovered that this was because of difference in their structural anatomy.

Lieberman then drew attention to the structural (anatomical) changes in the articulatory system which were responsible for the production of phonetary segments such as vowels and consonants. He however, paid no attention to, how did we evolve our ability to organize the movement sequence, and which vocal symbol to be uttered for which object or action? The structural changes which helped for the production of speech were not specifically evolved for that purpose but they were there as a result of our gradually acquiring erect bipedal position. Some structures were there evolved even millions of years before in our ancestors apes. Lieberman in his book "Speech physiology, speech perception and acoustic phonetics." has explained in detail; about the articulatory system and it is functioning about which we have already seen in Chapter 3. In detail, Lieberman said in order to explore language evolution, you have to completely abandon the idea that humans are born with some kind of grammar device. It was just not possible for both Darwin and Chomsky to be right and compatible.

Lieberman argues that not only you should study language evolution but also you cannot even begin to understand language if you do not start with evolutionary biology. Here he quotes famous evolutionary biologist, Theodore Dobzhansky, who says nothing in biology makes sense except in light of evolution. What he was quoting for this, is the Darwinian mechanism of channeling of a function that evolved for one purpose and used for another. What Darwin calls 'Descent with Modification' really it is exaptation of motor action.

When you start of thinking about evolution of language you have to think of bodily motor action which were evolved earlier and then parts that allowed us to order thoughts and cognition to put into words and speech. When it comes of sequential (syntactic) motor activity he thinks of motor system or movements. In case of language articulation he started with basal ganglion. The basal ganglion such as globus pallidus, striatum and other, are evolved earlier. Lieberman thought that they are responsible for learning patterns of playing tennis, dancing with sequential steps, and eventually expressing thoughts in sequential phonemes the words or speech. What induced Lieberman to think so was his observation of patients of Parkinson's disease? However, though the control of sequential motor action movements in general and also of fine movements such as playing tennis, playing piano, etc. might have remained with basal ganglia, after the evolution of neocortex the bodily motor activity including articulatory movements are shifted to precentral gyrus of frontal lobe and fine movements like playing piano to supramarginal gyrus in IPL. What he thought of fine movements might be true in animals in which neocortex was not developed?

Basal ganglia are responsible for learning patterns of motor activity. He argued. After his observation of patients suffering from Parkinson's disease, in addition to their physical symptoms, they had some speech defects such as:

1. To produce sentences that were particularly short, with only simple syntax.
2. When they were shown three pictures, and then asked to identify a picture that corresponded with sentence they had heard. They made errors, could not coordinate spatial and temporal sensations. They struggled with sentence.
3. They had trouble with 'active' and 'passive' voice. If they heard 'I ate a mango' and then asked 'who was the mango eaten by.' They had trouble with syntax though they understood the meaning.

Steven Pinker agreed on the subject of syntax and basal ganglion relationship and corroborated it with his own data. He added they had more trouble with past tense of regular than irregular verbs (like patients having gene FOXP2 deficiency.) The regular verbs with past tense have not to be memorized, they can just crank on it with recursive grammar as it cones sequentially. The basal ganglion disturbance in Parkinson's disturbs the sequential utterance of just automatically adding 'ed.' The irregular verb, however, is treated as a separate word.

4. They have also got difficulty with voiced and unvoiced phonemes. They have difficulty in following the sequence of utterance and voicing. This defect occurs along with syntactic errors.

5. They have got delay in comprehension of simple sentences and to create meaningful sentences.

One of the important function of basal ganglion is to maintain the thought and motor sequence.

Lieberman says we are using neural system which evolved millions of years ago for reasons other than formation of syntax, therefore he says the foundation of our syntactic ability, is an adaptation of our motor system, a primitive part of our anatomy.

In 2002, Lieberman says although our knowledge is at best incomplete, it is clear that many cortical areas other than Broca's and Wernicke's areas and part of the subcortical structural form, are involved in speech production. Part of the circuits implicated in the acquisition of the motor and cognitive pattern generators that underlie speech production, perception and syntax are some of them. He also includes in the list cerebellum, the prefrontal cortex, frontal cortex and anterior cingulate gyrus, and region of the brain associated with the visual perception and motor control. (Was IPL in his mind?)

Apes have larger area 44 of Brodmann, what purpose it serves? Cantapula and Hopkins suggest that apes are controlling gestures with this part of the brain (future Broca) in language like way. Human evolved the ability to perform gestures to visual perception. An ability that underpins verbal and gestural communication (through auditory system, refer Chapter 10). If a monkey hears breaking a shell, some of the audiovisual mirror neurons are fired. Visual sensation was already there when he had seen shell breaking. This is a long way from speech, but it does show that mirror neurons can link visual and auditory inputs. We have explained in our theory how the advantage of linking visual sensation to auditory inputs is taken in advantage of changing the auditory wave, exactly matching the visual inputs? So some basic mechanisms for grounding the evolution of speech analysis were, presumably already in place in the brains of our common ancestors who lived 20 million years ago. That is why? When he started engaging with the subject of language, he wrote of it as not so much as a new thing that humans have, we do it with a collection of neural parts that have long been available to us. Therefore he called language part primitive and part-derived just like Charles Darwin, who wrote in 'The Descent of Man' that language was half art and half instinct.

No one mutation of genes caused language to erupt from the mouths of our ancestors. Both he and Pinker agree that you have to start with the evolution to get really at the true nature of the language.

Nature *vs* Nurture

The behaviorist in the first half of the 20th century felt that psychology should be restricted to the study of observable stimulus and response relationship. But to what extent the rules of language are innate and to what extent they come as response to stimulus from environment and culture in early life was a question. The nature versus nurture debate about acquisition of language was very bitterly discussed. Everyone will agree that the words are not hardwired. Everyday, new words are added to the language depending on new cultural advent and progress in all fields of sciences. And they certainly serve as stimulus for response from the listener. Regarding the rules of language there is no such agreement. There are many opinions.

First, are the rules hard-wired and present from the birth? And a language switch is to be turned to turn the mechanism on.

The second view asserts that the rules are extracted statistically by listening a language. In this respect the work of Jenny Saffron is interesting. She says infants are statisticians by birth. They pick up often repeated clusters of words and rules, and learn, gradually the words and rules of language, from infant directed speech. This may apply to learning any new language around the world, but not to how rules evolved. This is somewhat like stimulus-response theory of behaviorist.

For this one needs to note the view of serial ordered stimulus-response (S-R) behavior that had prevailed since the inception of behaviorism by John Watson (1913), the founder of the behaviorism. It was his claim that this behavior of acquisition of language was produced by stimulus-response (S-R) arrangement of 'chains of reflexes in which the performance of each element of the series provides the excitation for the next. Skinner another proponent of S-R theory cited that just as a rat can be conditioned to press a particular lever to get a reward of food, an individual learns the benefit of uttering speech sounds and he must have learnt the language this way. When a small child utters a word, he gets a pat from the mother as a reward, and the child learns with this process of response rewarded and thus language is learnt by S-R and reward and punishment theory (operant conditioned reflex).

Question was production of speech when there is no external stimulus. Watson proposes that stimulus comes from thought. Thought was simply talking to oneself. This arrangement is of a chain of reflexes in which the performance of each element of the series provides the excitation for the next. Lashlay's argument against the chain reflex view of acquiring language, was, that there was not enough time, for feedback from the previous stimulus in fast serial response in speech. Bruce (1994) has cited quoting recent research that proves Lashlay wrong. He has pointed out that sensorimotor linkages can work fast enough. To this Lashlay's argument remains uncontested, that the choice of the subsequent response cannot be determined by, the previous stimulus-response, because a particular response is not an isolated event but is preceded by different stimulus-responses on different occasions. Lashlay therefore wondered, where the order for output comes from. He concluded that it comes from a level which is not only independent of output units but also of the input and thought structure. He noted the bilinguals can express the same thought with two different sentences structuring in two different languages. This requires a third level between the input, thought, and observed output, can it be mental level? This disproves the theory of behaviorist that only S-R chain reflexes sequence is enough to explain the origin of speech.

Teaching Language to Chimpanzee

Many experiments are undertaken by David and Ann James Premack and many others to teach language to chimpanzee and other pets. They do not learn language even when they are raised in human households. They still remain apes without language. Apes did not learn language not only because they had no Language Acquisition Device (LAD) (as proposed by Chomsky) but their anatomical configuration and neural development was also not competent for acquiring language.

Language Acquisition Device (LAD)

According to another view, the competence to acquire the rules is innate, but the exposure is required to pick up actual rules. This competence is bestowed on us by a still unidentified LAD, presence of which was originally proposed by Chomsky. Dr Ramachandran's view is, human have this LAD but apes lack it. Dr Ramachandran has given another example in support of existence of LAD. He proposes appearance of 'Pidgin' and 'Creole' languages, in the world as a compelling evidence of LAD. However this is not true.

The story is different. Labors were taken forcibly from Africa to Hawaii islands and Southern states of America during the days when labor trade was existent. They belonged to different tribes in Africa and they had their own mother tongues. They were then divided and put in different places. Perhaps the farm owners were aware of the story of tower of Babel. The labor's coming from different tribes did not know each others tongue except a few common words, leading to extremely limited vocabulary to communicate with each other. Their language was called as 'Pidgin' language. Dr Ramachandran calls it a pseudo-language—with limited vocabulary, rudimentary syntax, and a little flexibility. He further says that the next

generation who grew up surrounded by a Pidgin spontaneously turn it into a *Creole*, which is a full-fledged language.
He presents this fact as an evidence for the presence of LAD
(brain tell-tale). He himself has raised the questions how
language competence or ability to acquire language evolved
so quickly? How did it all got started? Though he calls Pidgin
as pseudolanguage it was a language of communication of
tribal groups and their languages were already present. This
itself proves that presence of LAD was not responsible in any
way for the evolution of language. The presence of LAD was
also not proved by its physical presence any time (brain tell-
tale).

Even if we call the process as genetic the motor action
potentials for the emergence of Pidgin vocabulary must have
developed first, which proves that presence of LAD was not
essential for the evolution of Pidgin or Creole language. Derek
Bickerton who has worked in Hawaii with his colleagues is
not sure about the origin of Creole as an evidence for the
presence of LAD (Fig. 7.2). In the map that he has given with
his article in "Emergence of languages" he has shown that
Creole in every state or an island is different and has called it
as Seychelles' Creole (French), Hawaiian Creole (English),
Angolan Creole (Portuguese) and so on. Which obviously
means children developed the Creole, based on different
languages with which they came in contact, which was the
language of the farm owners of their ancestors. They developed
Creoles borrowing the syntactical and other aspects of the
language from the languages of their farm owners. This process

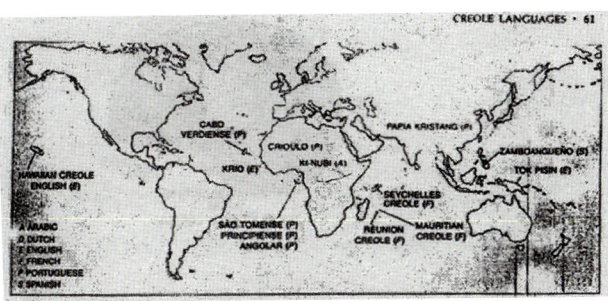

Fig. 7.2: Evolution of Creole languages

must have taken more than a few generations. Bickerton does not conclude that Creole originated from Pidgin because of LAD. Bickerton in the end says "How did human language originate? What are the minimum prerequisites for such a thing as language to arise in a species? If such questions can be answered we shall be much closer to understanding what makes the human species different from others?" He dose not make a mention of LAD, perhaps because he was not aware of it and was not aware of the structural (anatomical) and neurological advances which had taken place in human species, which were responsible for origin of speech and language. Perhaps what Bickerton was missing is what Lashlay had wondered, from where came the orders? we have already seen in detail about this in the discussion of nature *vs* nurture.

The theory that I am going to present in this treatise—of course based in the light of natural selection—is based on these advances. And try to answer the difficulty of intermediate level raised by Lashlay.

8 How Speech Started Origin of First Word?

We have discussed the theories of, origin of language, proposed by various thinkers, most of them of course were linguists. But as we have already seen language is not a one package. The word has to be organized first in the brain pairing to the object or event, and then order must be received by the articulatory system for its utterance. This is observable output of language. Is phonology still a mystery? Syntax is a differentiating factor between a protolanguage and modern human language. Phonology is the key to understand the aspects of transition from protoluanguage to language. Various linguists and other scientist have put forth various opinions in this respect of utterance of **first word.** Let us review some of them before coming to the theory of the author of this treatise.

Human words are much more than just a sequence of sounds. They refer to something in the world an object or an event. A word is an arbitrary association between sound and meaning. There is nothing in the sound of a word that tells you what it means or what it does not. You must learn it. Whenever you hear a word you know what (if anything) it means if you have heard it previously with reference to an object or event? However, you know that some words stand alone like hello, ouch and yes. And others can join together to create words like 'heretofore' and 'bedroom'.

THEORY OF PETER MACNEILAGE

Basically human produces vocalization the way the other mammals do—by producing sounds using parts of the body that initially evolved for feeding and breathing. We even use the same three systems: Respiratory, phonetary and articulatory for voice production. And we use them in a similar way. The main difference is the simultaneous generation and modulation of biphasic cycles. What are these biphasic cycles? They are sequences of two alternating movements, in each of the three systems. In respiration the biphasic cycle is the inspiration and expiration. Expiration gives energy to speech in the form of air released out of the glottal opening in puffs. In the phonetary system the biphasic cycle—the dominant contribution of this system in all mammals—the vocal folds alternating regularly between open and close positions resulting in voicing. Our pre-human ancestors were already capable of abducting the vocal folds at the termination of vocalization and closing or adducting them for protection of respiratory tract from foreign material entering in. In nonhuman mammals the articulatory system is typically employed only in an open configuration during call production. Here the biphasic cycle is very basic and in human the biphasic cycle of articulation is the opening and closing of the jaw. Thus, in human a series of open close alterations is the mode. Indeed as we have seen, such a regular alteration between a relatively open and relatively closed configuration—closed for consonants, open for vowels is basic enough to be a definitional characteristic for phoneme production. The third level of cyclisity in human is opening and closing of mandible or jaw. This in mammals is the third level of cyclicality, which is articulatory cyclicality. What was the position in other pre-human mammals? There was inability to produce vowels and consonants in them, due to anatomical constraints. Open and close cyclicality of mandible was accepted by MacNeilage as third cyclicality. Though it was not used in other mammals for voicing, it was present for feeding, drinking–sucking for at least million years. These biphasic cyclicalities were used by pre-human species, when just by chance they were accompanied by phonation with tongue smacks, lip smacks or teeth chatters they can be considered as protosyllabic or later protolinguistic utterances. This is an example of exaptation—the term introduced by Gould and Verbs (1982) which literally means borrowing the activity of one function to another function, here from the realms of ingestion to phonation.

Lund and Kolta (2006) recently proposed that the brainstem circuits which control the activity of chewing or mastication control the speech, though with different system of coordination of muscles. This similarity suggests that processes related to chewing were indeed co-opted for speech. The possibility that the modern man may have exapted the mandibular cycle for speech, makes it unlikely, that cycle of speech was constructed from, scratch. Then why a little attention was given to the fact, that ingestive cyclicality was precursor to speech, he asks?

The animal calls are a few in number and are holistic. However, at least some calls have harmonic properties, indicating the presence of vocal fold vibrations analogous to that found in humans, others are noisy, indicating some turbulent source presumably at the level of glottis. (Can this be a precursor to voiceless sounds of some consonants?—Author's opinion.) They combine noisy and harmonic characteristics. Repetitions of the same sound is the mode in primate calls. They have no ability to produce different meanings with different mandibular cyclicalities. These are relatively rare in non-human vocalization systems. However, visuofacial communicative gestures are extremely common. Perhaps these communicative events from ingestive cyclicality and its exaptation for speech by our ancestors may have been the evolution of protosyllable and protolanguage.

MacNeilage says even though the lip smacks are unlike speech not accompanied by vocalization, its surprising that more attention has not been drawn to the similarity between the movements of mandible (mandibular cyclicality) lip smacks, and the production of a syllable. He thinks lip-smacks could be a precursor of speech. He gives following reasons to support his claim. (i) Lip-smacks occur very frequently, (ii) it occurs in the context of social interaction, (iii) it is one to one reaction sometimes followed by listeners response, and (iv) finally it might be accompanied by phonation. We will later review whether his view is sufficient to accept the lip smacks, tongue smacks and teeth chatters, accompanied with mandibular cyclicity as precursors of speech. What he has neglected though is the main role of tongue and the mouth organs in production of first word? Perhaps because these movements are invisible.

MacNeilage's theory therefore remains incomplete. He has also missed the link, who orders for the action potentials for the utterance.

Dunbar's (1996) Hypothesis

Speech might have first evolved in the form of vocal grooming. In primates physical grooming is very much used to keep social bonding. With increasing size of a group and emergence of new species. It might be necessary to spend more than 40% of their time in physical grooming, and that became too much time consuming. Larger groups depended more upon economical and less, paired centered, alternative to physical grooming; and this in Dunbar's term, was verbal communication. Action frames or motor frames accompanied by phonation were essential for communication. To summarize the sequence of four events must have given us the first word. (i) Grooming with lip smacking, (ii) phonating with lip smacking, (iii) increasingly substituting, smacking together with phonation, (iv) adding specific semantic information to this vocal components.

Donald—the Evolution by Mimesis

Merlin Donald (1991) incorporated the role of motor (action) side of humans in their mental evolution. In addition, she correctly stresses the sociocultural bases of hominid evolution without which there would have been no language at all. Languages are results of social agreements. She insists that to make the sound pattern of first words of a language there was a necessary precondition of evolution of protolanguage (Donald 1999). In her view the ability to make forms for signaling the words cannot be just taken for granted as automatically following from a higher order, word making capacity. What was needed in advance was a species wide representational capacity to form symbols (words) which stood for concepts. Any language is a set of conversations shared by sender and receiver for transmitting symbolic information in the form of gestures and sounds. All gestures and intentional vocalizations are ultimately actions of musculatures and in order to generate greater varieties of gestures, motor behavior must have become plastic and less stereotyped. To vary or refine any action one must carry out a sequence of cognitive operations. Such as (i) rehearse the action, (ii) observe the consequences, (iii) alter the form of original action, varying one or two parameters, detected by the memory of previous consequences, or the idealized image of the outcome "A rehearsal loop." Human infants do this type of things very often especially in infant babbling. This capacity is not seen in great apes. It is uniquely human and forms the background for human culture and language. The advancement of motor skills was generalized adaptation for all actions such as singing, music, etc. These developments are uniquely human. It is evident in our mastery of motor rhythm. We can produce a rhythmic pattern not only vocally but also with our feet, fingers or whole body.

Donald call the basic capacity discussed as "mimesis" which is a capacity to mime or reenact events. In Donald's opinion this capacity evolved sometimes in the course of evolution of 'Homo erectus'. How mimesis might have developed by natural selection? Donald's answer was, it has been an adaptive response to pressures for social communication to maintain sociocultural solidarity. She considers all these changes to be prelinguistic. But mimetic skill provided only an essential background for speech and language because that enable to link the words with sound structure. The other development necessary for protolanguage to occur was the capability of forming conceptual side of words which was not there.

We have examined the views of MacNeilage, Dunbar and Donald. All of them have given importance to visual movements such as mandibular oscillations and nonvocal sounds such as lip smacks, tongue smacks and teeth chatters. Perhaps it was not their field to give more importance to tongue movements and some definitive importance to the physical properties of the utterances. Phonemes acquire symbolic value after they are uttered with intention. The sound accompanied automatically with bodily organs such as tongue, teeth, or lips are not intentional, and therefore, they have got no speech and language value. They can be sounds accompanying speech and language or just happen to be produced without any intention.

Speech and language, however, has got the quality that it is intentional exchange of information. Dunbar has proposed the value of grooming as precursor of speech. But words cannot originate from grooming as grooming has got different quality every time depending on the subject to whom it is directed and the occasion. It will not have the sound subject pairing required to get the status of word. Donald's imitation or mimetic hypothesis does not tell us how you imitate the objects producing no sound and achieve sound pairing required for the utterance to become true speech? All these hypotheses cannot make us wise about the origin of the first word. Did it come from gestures as mentioned above? The hands spread and moved up and down for the sign of flying bird flipping. This natural relation is called 'Iconicity' but how it initiated proper word?

Disappointingly, neither the orthodoxy nor the mavericks had much to say about actual origin of the stage of pairing sound with meaning. None of the theories mentioned above MacNeilage's based on rhythmic movements or cyclicities, Dunbar's based on vocal grooming or Donald's mimetic principal could do it.

Steven Mithen's Theory

Steven Mithen, Professor and Head, Department of Anthropology and Environmental Sciences. He has proposed a thought about origin of speech in his book 'Singing Neanderthals.'

Chimpanzee was communicating with the members of his tribe by vocal calls called barks, grunts, etc. They made use of gestures along with the vocal calls such as facial movements, hand movements and signs. Even birds communicate with their singing voice. It is logical therefore to accept that pre-human species must be communicating with vocal messages. The vocal calls do not leave behind any direct evidence or fossils of language with which they might have communicated. We can certainly take it for granted that they were also communicating with vocal sounds. Their calls were limited. Linguist in Cardiff university Alison Wray has called these vocal calls as 'holistic'.

Holistic means each call has only one meaning. There is no consonant—vowel sequence in the vocal calls. Therefore there are no words and unlimited vocabulary. These calls are produced according to the needs of lifestyle and needs for survival and reproduction. Their needs were limited. Steven Mithen has coined an acronym **Hmmmm** for these calls. **H** stands for holistic, first **m** stands for manipulative that is for giving danger calls to other members in the group about predators and natural calamities, the second **m** stands for multimodal capacity that is the call is composed of intentional vocal sounds and imitation of other environmental sound, noises, etc. The third **m** represents mimesis that is imitation of sounds in nature and surroundings. The last one is for musical. This might be rhythmic sound with variations in pitch and intensity to attract the mate for reproduction and to use at the time of group dancing. Let us also call this vocal call as Hmmmm. This call has no symbolic relationship with any object or event. But the call has gestural support. Our today's language also has got gestural accompaniment. Some say that we support our talk with about 65% gestural signs. Some of them are very popular, a sign of V for victory. A sign of thumps up, not only for popular cold drink but also for all cold drinks and for many other purposes. These vocal calls were limited for many reasons. (i) The groups of species was having a few members and not much of vocal communication needed, (ii) their cognitive development was of single domain and there were not many subjects to talk about, (iii) they had very limited collective activities as they were not constructing huts, using fires, stitching clothes, etc., (iv) while manufacturing stone tools they were siting away from each other. There are evidences for that. There was, therefore, very limited need for communication, (v) because the calls were accompanied with gestures more than one meanings could be suggested from one call, (vi) unable to learn new calls in the wild, (vii) anatomical constraints.

Alison Wray says that the pre-human calls were holistic and manipulative rather than referential and compositional. She further suggested that the calls could be segmented and could acquire different meaning for each piece, therefore she called them as precursors of language or protolanguage. Her views are diametrically opposed to those of Derek Bickerton. He thought words without grammar as protolanguage. Social complexities and increasing number, acting as selection pressure for enhanced communication was the stimulus for their utterances. These were possible by the physiological changes associated with bipedalism, which enabled the *Homo erectus*, a significantly larger array of vocalization than either their immediate ancestors or modern day apes. It is quite legitimate to refer to holistic vocalizations of *Homo erectus* communication system as protolanguage was his opinion. But this protolanguage with increased number of holistic vocalizations gave no foundation for word without grammar, type of protolanguage as proposed by Bikerton. While this type of protolanguage of Bickerton, might have been adequate for communicating some basic observations about the world, they would have been unsuitable for expressing, thought for, what Alison Wray describes, as the 'other kind of messages'—those relating to physical, emotional and perceptual manipulations? Wray's concept of protolanguage has come under heavy criticism by Bickerton and Maggie Tallerman a linguist from university of Durham. They argue that because a holistic protolanguage of Alison Wray is evidently an evolutionary continuity with ape like vocalization it simply cannot be a precursor, of protolanguage to language of modern human. Tallerman shows 5 distinctive characters between human speech and primate vocalizations: (i) speech and vocalizations are handled by different parts in the brain. Vocalization by limbic system in right brain and speech by Broca's area in left, (ii) primate calls are involuntary, (iii) both have different physiological basis, (iv) primate calls are genetically transmitted, (v) human speech has got dissociation in sound and meaning, which is absent from primate calls.

Another criticism was that the holistic calls would have been too short, too ambiguous and too a few in number to have constituted a feasible protolanguage from segmentation of calls proposed by Alison Wray.

Bickerton's protolanguage was problematic because if it had evolved during the period of *Homo erectus*, it remains unclear, why such a protolanguage existed, since 1.8 million years BP without rapidly evolving into a full modern language prior to 200,000 years which was the period of appearance of *Homo sapiens*?

Bickerton proposes three types of linguistic fossil: (i) the Pidgin language, (ii) language of children bellow two, (iii) out put of language trained apes.

To accept any protolanguage and that evolving into perfect language we first need to justify the presence of language ready brain, which initially produced something different from and less complex than today's language. We find single words and not syntactic structures in that type of language. Now that we have accepted syntax free view of protolanguage, a shift from internal concepts directly to external linguistic expressions and subsequent emergence of modern language, then we will also have to account for, from where words came? In such a case, you are in a bad place trying to motivate a lexical protolanguage which never existed.

Musical protolanguage is an attractive proposition since it provides an intermediate stage which allows the gradual development of many of the aspects of physiological and neurological structures required by the modern language. It should also justify the problem of manual signing and its transition to vocal calling leading to excessive social grooming while working. In this model music is the oldest communication system that pre-dates modern language. While there might be continuous discussion over the dating of this transition there is growing consciousness about the key role played by the development of the complex synthetic phonology in place of the emotionally linked holistic phrasing signals of early musical protolanguage such as dance, singing or otherwise.

Mithen Hmmmm language an attractive proposition for musical protolanguage, since it effectively provides an intermediate stage which gradually allows development of the many of the neurological and structural changes required by modern language. Mithen sees break up of this holistic protolanguage stage as relatively late event in evolution, and one which distinguishes our own species from the Neanderthals.

Pinker dismissed music as auditory cheesecake in other words music is a parasitic system (or spandrel) perhaps relying on the neural developments for language and originating along with language subsequently.

We have already seen the Alison Wray's idea that the holistic calls were manipulative rather than compositional or referential. How then Hmmmm provides as evolutionary precursor of language? Wray describes a process for which she uses the term segmentation where the holistic phrases began to break up into separate units, each of which had its own meaning. It could recombine with units from other utterances to create an infinite array of new utterances. This is the emergence of compositionality. How segmented units got meaning? Wray suggests that it may have arisen from 'chance association' between the phonetic segment of the holistic utterance and the object or event to which it is related.

Bickerton and Tallerman, however have questioned the feasibility of segmentation and likelihood of any chance association.

The presence of onomatopoeia, vocal imitation and sound synesthesia would have created a non-arbitrary association between phonetic segments of holistic utterances and certain entities in the world. This is better explanation than the chance association concept of Wray. Problem remains why did Hmmmm utterances of *Homo erectus* and Heidelbergensis remained intact without segmenting. The reasons might be that the social life of these groups was such that there was no need to them for more novel utterances needed for compositional language. There were a few or no opportunities for mixing with other groups. With emergence of *Homo sapiens* such opportunities did increase, as a result of the development of economic stability and culture. To exchange messages and exchange relations between the groups, evolution of compositional language was a consequence. Or, was it the reverse? Or, was there a strong and mutual feedback between the two?

Another thought is that kick-start for such developments may have been a chanced genetic mutation. That might be second possible means why segmentation of Hmmmm occurred in Africa just before the evolution of modern human. This may have provided an opportunity to identify the phonetic segments (? or words) in holistic utterance. This capacity might have been possible because of some genetic mutation. Chance mutation of the genes was responsible for the gene FOXP2 appearing in *Homo sapiens* in Africa. There may have been the other genetic mutations also at about similar date that enable the transition of holistic to compositional language perhaps through the appearance of a general purpose statistical learning ability. How compositional language evolved from holistic phrase? The compositional utterances that emerged from holistic phrases by a process of segmentation would have begun as a supplement to the Hmmmm communication. Indeed, the transition from Hmmmm to compositional language would have taken many millennia, the holistic utterances providing a cultural scaffold to the gradual adaptation of words and new utterances structured by grammatical rules. Moreover, Mithen says rather than means of communication the first words might be of significance for self talk. Communication, could have continued by Hmmmm for some time. Compositional language would then have become a supplement to Hmmmm and eventually, the dominant form of communication, for its greater effectiveness of transmitting information.

Thus, in spite of the idea of Steven Mithen, of holistic vocal calls and their segmentation somewhat similar to phonetic elements or words ultimately he had to take resort to chance genetic mutation and the work of Jenny Saffron. But could not explain how a word appeared in the holistic segment for statistical extraction. He had also to resort to appearance of language gene FOXP2 and other genes for gradual transition of holistic vocal calls and their segments to compositional language. There is no logical explanation also, for sound-object or sound-action pairing. These were the major defects in Mithen's theory.

In Chapter 10, we will see whether any other logical thought can be provided for the origin of speech.

9 Dr VS Ramachandran's Hypothesis

Dr VS Ramachandran is the Director for the Center for Brain and Cognition, Professor in the Department of Neurosciences program at the University of California, San Diego and Adjunct Professor of Biology at the Salk Institute for Biological studies. He has proposed his hypothesis of evolution of speech and language in his book 'The tell tale brain.' In his book he has taken a review of research of 'mirror neurons' and their function in the origin of language. He has proposed his theory of synesthesia and synkynesia. We will take a review of it and some shortfalls of the hypothesis.

I have given a more space for discussing Dr Ramachandran's hypothesis because as far as I know he is the only author from medical field, who has given a thought to the subject of origin of speech. He is moreover the first to think that the object or event for which a word is used must have some non-arbitrary relation between the visual profile of the object or action and the word that is used as a symbol for that, which also forms the basis of my theory.

We have already seen the anatomy of brain in Chapter 3. We know about the IPL or inferior parietal lobule, or the lower and backward part of the parietal lobe. It plays a major role in the origin of speech. In the last chapter a review of various theories proposed by thinkers about 'the origin of speech and language' was taken. None of the theories could satisfy the evolutionary process behind the origin of speech in *Homo sapiens*. In this chapter we are going to consider hypothesis by Dr VS Ramachandran. Before going into the details of the hypothesis of Dr Ramachandran and discussing its merits and demerits, let us have a look into the evolution of brain structures and functions taking place in the human brain as a part of the evolutionary process. We have already seen the details of the anatomy of brain in Chapter 3. The hominine brain reached its present size 200,000 years BP. However, the art of construction of fire place or a dwelling hut, etc. evolved only 75,000 years BP. Dr Ramachandran adds language to that. However, for the development of all these skills, language evolution was necessary prior to that. Sophisticated tools evolved with *Homo erectus* 1.8 million years BP. These were called levollois tools. They were making use of fire but not fireplace. No other signs of cultural development were there. There was no speech and language as for the evolution of speech, the evolution of SVT, which is competent for the production of all the different formant frequencies for phonemes is required. The utterance of different phonemes of language had not taken place.

IPL AND ITS ROLE IN SPEECH PRODUCTION

IPL or the inferior parietal lobule plays a major role in the origin of speech. In lower mammals the IPL is not very large. It becomes more conspicuous in primates. It is disproportionately large in the great apes. It reaches its climax in modern human. In human and humans only, it splits into two parts. The two parts are named as, angular gyrus, and supramarginal gyrus. Splitting into two, indicates that something more is going to happen in this region. This region is a crossroad among vision, represented by angular gyrus of occipital lobe, touch and position of muscles and joint sense by parietal lobe, hearing by Wernicke's area of temporal lobe. Vision and hearing or auditory sense are two different modalities. What happens at this crossroad of different sensations, is to dissolve the barriers of different modalities and initiate cross modal abstraction or adaptation? This is the function of mirror neurons or circuits. These mirror neurons are plentiful in IPL region. Before going into the details of what exactly happens in this region let us know something more about these mirror neurons which play a very important part in the evolution of speech (Fig. 9.1).

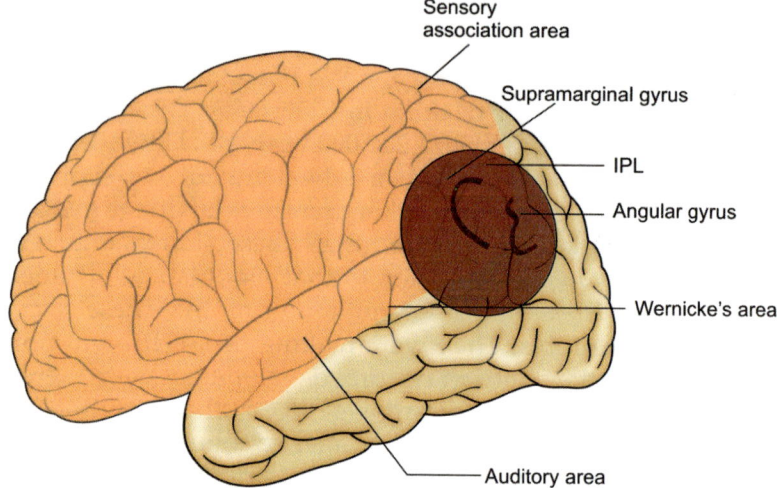

Fig. 9.1: Inferior parietal lobule

MIRROR NEURONS

Humans are called Machiavellian primates (having skills in using evil means to gain one's aims). For this, skill, predicting other man's mind and outsmarting him was necessary. This is called 'theory of mind.' What neural circuits are responsible for achieving this? How we evolved them? Solutions to this riddle came from the discovery of mirror neurons. When we say mirror neurons it actually means neural circuits.

In 1990, Giacomo Rizzolatti of university of Perma in Italy, reported some of the neurons were firing, when an ape was not performing an action itself, but was observing someone else in that action, this was interpreted by him, that they were perhaps adapting other animal's point of view. He named these neurons as 'mirror neurons.' In human, and humans only they have become sophisticated enough to interpret even complex intentions of someone else. They adopt someone else's conceptual vantage point "I will try to see it from your point of view" was a magic step from literal to conceptual. This is known as 'theory of mind'.

The mirror neurons enable us to mime (imitate) the lip, tongue movements of others which may produce evolutionary basis for verbal utterances. Language has not remained mysterious and does not need the presence of 'language evolution', it requires many specialized areas in the brain. Left IPL in parietal lobe is one of them. It is crucially involved in the evolution of speech especially in the representation of word meaning. This area is very rich in mirror neurons. Is there any supporting evidence for this? Yes. If anytime you have to make judgment about else's intentions, you have to run a virtual reality simulation of the corresponding movements in your own brain. You cannot do this without mirror neurons.

If you are sitting in the company of your friends and one of them wants to drink a glass of water, gets it and drinks it, you also make all the movements of having a glass, taking it to your lips, drink, even though you have got no glass in your hands. These are all dummy movements in your brain. You cannot do this without mirror neurons.

Since Rizzolatti's research another type of mirror neurons have been found in anterior cingulate gyrus a part of limbic system (Fig. 9.2). Limbic system is supposed to be a seat of emotional responses, including seat for pain from primary pain neurons. During surgery of brain, under local anesthesia, it was found that if patient could watch someone else being poked, sensory neurons of the patient were equally vigorously responding or neurons were empathizing with someone else's pain. This

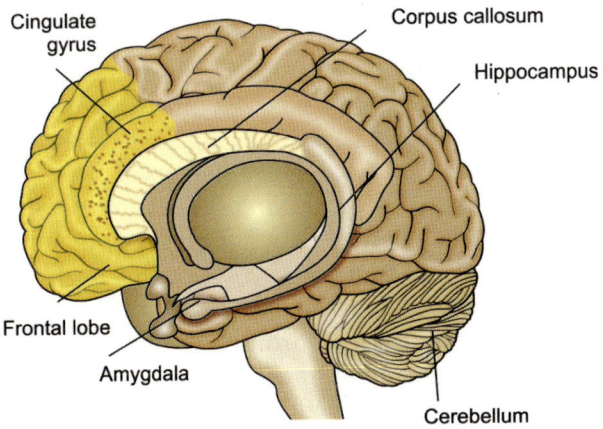

Fig. 9.2: Limbic system

response is possible because of presence of mirror neurons. Dr Ramachandran aptly called them 'Gandhi neurons'.

Here I remember an event in the life of great exponent of Indian classical music, "Kumar Gandharva." Once he was a witness to a sight of slaughtering of a she goat, when he looked eye to eye in her eyes, he saw, only the worry about the future of her a little lamb. The great empathy gave birth to a great ragdari (symphony). This might be the effect of mirror neurons present in cingulate gyrus of limbic system.

But I wondered what about this another event?

In some countries still there is a custom of a capital punishment in the form of public lynching for certain crimes. The public gathered around, is seen enjoying the event and jeering. What kind of empathy this can generate and what type of mirror neurons might be involved in this?

One more question arises what prevents you from blindly imitating every action you see or literally feeling someone else's pain by gestures? Frontal inhibitory circuits that suppress the automatic mimicry,, when it is not appropriate to express these by action suppress this. Another question is, whether mirror neuron function is innate or learnt, or a little of both?

Andrew Meltzoff notes his observations in this respect. Newborn infant protrudes the tongue or smiles while seeing his mother protruding the tongue or smiling. This cannot be based on learning. It has to be innate and hard-wired. Infant cannot see its own face and it is too early to learn the mimes for smiling.

Translation of maps is what precisely mirror neurons do? If there is a glass cracked or a paper is, being cut by scissors, what mirror neurons do is, translation of brain maps in one dimension that is visual, to Kiki sound of cutting or jaggedness of broken glass, in another dimension of sound or auditory maps?

The less obvious function of mirror neurons is abstraction, something in which humans are especially good at. The main computation done by mirror neurons is to transform a map in one sense such as visual appearance of an object, big or small, tall or dwarf or some other action fast or slow, into another dimension such as auditory in the observer's brain. **Your brain**

is performing an impressive feat of abstraction in linking the two, visual and auditory maps. The two inputs are entirely dissimilar, your brain homes in, on their **common denominator** very swiftly when you are asked to pair them up. Dr Ramachandran calls this process as **cross-modal-abstraction.** This ability to compute similarities despite of surface differences may have paved way for more complex type of abstractions. This is possible probably only in modern human species. Mirror neurons may be the evolutionary conduit that allowed this to happen. If this is correct, some of the functions of mirror neurons may indeed be acquired through learning, on the basis of a genetically based scaffolding, unique to human only.

The cross-modal-abstraction such as, sensory to sensory, called synesthesia, or sensory to motor, or motor to motor, called synkinesia freed us from genetics and therefore from Darwinian shackle of acquiring new faculty only via variation and therefore by natural selection. You can do it by our ability to learn from one another by imitating or reading others mind, which is a function of mirror neurons. Brain allowed us rapid spread of our inventions such as using fire place, constructing houses, and new tools and even evolving new words. Mirror neurons played a very crucial role in this.

DR RAMACHANDRAN'S HYPOTHESIS

Let us now look to the hypothesis of origin of speech and language as proposed by Dr Ramachandran. He says the question is not how speech and language evolved but how the language acquisition device (LAD) or the language acquisition competence or the ability to acquire language so quickly evolved? He thought, this competence is controlled by genes that were selected for it by the evolutionary process. Our question is, he says, why were these genes selected and how did this competence evolved? Is it modular? (Sequence or pathway that is designed for a function which can be repeated in similar function is called modular.) How did it all get started? And how did we make the evolutionary transition from grunts and howls of our ape-like ancestors to the transcendent lyricism of Tagore?

Here he recalls an experiment undertaken by him. He calls it a Bouba-ki-ki experiment. Bouba is the name of a tree having very big base, so much so that ten persons can stand around it hand in hand easily. It is an African tree, however, one can see some, on the Mandu fort near Indore in India. Ki-ki is the sound heard when a glass cracks or a paper is being cut with scissors.

He wants to know how the first word evolved among a band of our ancestors in the African savanna between one and two hundred thousand years ago? Since words chosen for the same object or event are often utterly different in different languages one is tempted to think that the words chosen for a particular object or event are entirely arbitrary. This in fact, is a standard view amongst linguists.

It cannot be that a few adults sat around a fire one night and decided let us call this object that is flying up in the sky 'a bird.' Dr Ramachandran says, thinking that way about, the origin of word is silly. Then how did it happen?

There must be some nonarbitrary built-in correspondence between the word and the object or action. It might be in visual shape and the sound or at least a kind of sound that might be its partner. This preexisting bias, he says may be hard-wired.

Here he does not explain what is the preexisting bias between these two entities which are entirely different from each other, and what was the reason for the bias being hard-wired. He thinks, it is in some way built in? He further says that this bias may be very small but sufficient to start the process.

How can be the bias be small? Is it because it is in respect of not all but only some similarities between the two? And what is the exact process, it starts to complete the nonarbitrariness amongst the two?

Here, my view is, contrary to what Dr Ramachandran says, that here the nonarbitrary (built in) correspondence between the word (phoneme, the term is not used by Dr Ramachandran) selected and the object or the event it is going to represent, is neither built in nor preexisting and hard-wired. In the presence of mirror neurons some dominant common denominator or similarity between the two is selected and the process of non-arbitrariness is started. (It is incomplete here in IPL. It is further processed to its complete nonarbitrariness in Broca's area. How this later process takes place, we will discuss further in the present author's hypothesis later?)

Dr Ramachandran's idea of built in or preexisting bias takes us quite near to the onomatopoeic theory. In this theory the sound directly refers to the object. For example sound meow-meow directly results in a child calling it a mow or in case of naming the crow from its call. The onomatopoeic theory posited that the sound associated with an object becomes shorthand to refer to the object itself. But he himself has refuted that this onomatopoeic theory is sufficient for its application for the utterance of a word. The onomatopoeic theory held that the link between the word and sound was arbitrary and merely occurred through the repeated association.

Here he proposes his theory of 'synesthesia'. This insists on having some physical parallel in between the two. But he does not explain what that parallel is, between the two different sensations, the visual sensation created by the object and the sound which synesthetise with it to create an auditory sensation. This auditory sensation created by synesthesia between the two is the first stage in the formation of the word. He says that the rounded visual shape of the Bouba tree does not make a rounded sound or any sound at all, but its visual profile resembles the profile of the undulating (vibrating) sound at an abstract mental level. Synesthetic theory says the link becomes nonarbitrary and grounded in a true resemblance of the two in a more abstract mental space.

He does not explain how the relation becomes nonarbitrary when they enter the abstract mental space. We really do not know yet what is a mental space. Perhaps what he wants to say is the visual sensation for the object on one hand and its translation to acoustic or auditory sensation on the other hand is effected by, cross-modal-abstraction in the presence of mirror neurons in IPL. But it seems that there is one shortcoming in this thinking. The cross-modal-abstraction (or adaptation) only acts according to whether the object is big or small and can select front or back vowel for the auditory wave which represents the parallel between the two, the object or action and the sound. But here there is nothing to decide or compare as to how much big or small is the object or how fast or slow is the action. There are more than one front and back vowels and therefore to abstract and develop exact nonarbitrary relation between the two something more has to happen. What is that something more, and where and how it happens, we will find out in Author's theory in the next chapter.

Sound, or auditory sensation which is parallel to object effected by synesthesia between the visual and sound sensation is carried to the Broca's area in frontal cortex. Further he says that this area contains maps for motor movements of lips, tongue, palate, etc. or in short articulatory movements for the production of sounds of phonemes (he has not used the word phoneme). This area is also rich in mirror neurons providing interface between the articulatory movements for production of sound, listening to sound and (least important) watching lip movements.

Dr Ramachandran says that there is nonarbitrary (exact) correspondence and cross activation between brain maps for sight and sound (think of Buoba-ki-ki experiment) or visual and auditory maps on one hand and the motor maps for gestural movements on the other in Broca's area. This may sound a bit cryptic but think of mouth and jaw movements when uttering a sound for small objects such as 'diminutive' or 'tinny', 'winny', he says, the mouth, lips and pharynx actually become small, and for large objects words like 'enormous' the mouth becomes physically open and enlarged. The abstraction device that translates the visual and auditory contours into vocal contours specified by muscle twitches act in the presence of mirror neurons.

Here my objection is, if we observe the words for small objects like tinny-winey, they contain mainly [i] as a vowel. For the utterance of vowel [i] the articulatory movements are different than what Dr Ramachandran has noted [i] is a front vowel and for the utterance of front vowel the lips are widened and are together? For the words like enormous for large objects the principle vowel is [u], as in hood, the vowel [u] is a back vowel and for the utterance of back vowels the lips are rounded and not opened and large (Figs 9.3a and b).

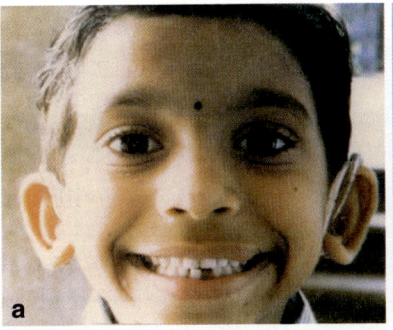 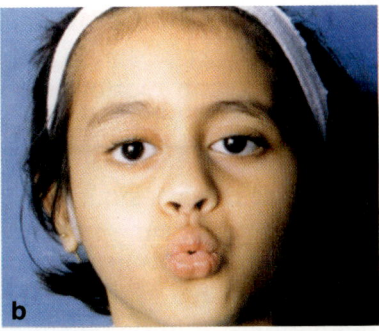

Figs 9.3a and b: (a) Lip movements for vowel [i]; (b) Lip movements for vowel [u]

Another objection is the stimulus that executes the muscle movements for the sound of a phoneme, do not stimulate the individual muscles through Broca's area. The stimulus that is given to the articulatory system for the production of sound of a phoneme is 'goal oriented'. The stimuli for individual muscles originate from the brainstem area. What is the advantage of stimulus being goal-oriented and not for individual muscle. We will see in more details in Author's theory in Chapter 9.

He has given the example of ki-ki experiment for his argument. If you are cutting with a scissors, it makes a ki-ki sound. He further states that since cortical areas concerned with the hand and mouth are next to each other perhaps there is actual spill over of signals from hand area to mouth area and there is clenching of jaws, as an unconsciously echoing response to hand movements. He considers this as gestural movement parallel to ki-ki sound. Perhaps there is actual spillover of signals from hand to mouth. As in synesthesia there appears to be a built in cross activation of brain maps, between two motor maps rather than between two sensory maps. The new name he has suggested is 'synkinesia.' This translation from hand movements to the clenching movement of mouth is not a direct response to auditory stimulus. Such responses are seen even when clenching takes place when auditory response is not involved. If someone is pushing a heavy weight, then also there is spill over to mouth area and as a response, related oro-facial gestural movements of clenching of mouth are seen. The response to ki-ki sound accompanied with cutting can be considered as an auditory sensation stimulating the articulatory system for the production of parallel sound or phoneme. There it will be a sensory to motor translation (or adaptation).

Further he says synkinesia may have played a pivotal role in transforming early gestural language of communication of primates into (oral) spoken language. We know, that vocal calls in primates are given by limbic system (Fig. 9.2) in right hemisphere. Limbic system is part of the mechanism for emotions. The anterior cingulate gyrus is part of this system and is responsible for the motor movements accompanying emotions. He says, if manual gestures were being echoed by orofacial movements while the creature was simultaneously making emotional utterances the net result would be what we call 'word.' In short ancient hominines had a built in, preexisting mechanism for spontaneously translating gestures into words. This makes it easier to see how a primitive gestural language could have evolved into speech—an idea that many classical psycholinguist find unappealing?

In my view, there is no preexisting mechanism for spontaneously turning gestural movements into oral movements for production of phonemes. Here I will say that there is a little confusion. The movements of mouth, lips and tongue at the time of production of a phoneme are not the movements accompanying emotional utterances. They are the movements of articulation of a phoneme. The motor movements which may accompany emotions, are the movements like perspiration, raising of hairs or involuntary ejaculation and these are executed by motor area in anterior cingulate gyrus, which is a part of associated motor area or area 6 of Brodmann going deep down to anterior cingulate gyrus medially in longitudinal fissure (Fig. 9.4). Therefore, there is no question of gestural movements being translated into movements of lips, tongue, etc. the other emotional movements affecting speech are part of suprasegmental and prosodic part of speech about which we will discuss later. Gestures have phylogenic importance as they are present in primates and pre-human species, and that was an important aspect of their communication system. However, in humans they perform a very important function in translating visual sensation into auditory sensation, a cross-modal-abstraction (Fig. 9.4).

This we will discuss further in my theory of evolution of speech.

Dr Ramachandran has stated that the auditory sensation that reaches the Broca's area stimulates the movements of lips, tongue and mouth as the motor areas for these movements are already there. However, as we have already seen, the stimuli, for the production of speech sounds, or the production of a phoneme are goal-oriented and stimuli for individual muscles are not given by Broca's area. The individual muscle stimulation comes from the brainstem area. What is the advantage of the stimulus being goal-oriented in this area, we will discuss later. Is there any proof for the statement that the stimulus for the production of articulatory movements from this area, is goal-oriented? Yes, the proof comes from the bite block experiments conducted by Folkins and Zimmerman (1981).

This experiment shows that the muscular activity of speakers producing utterances while their jaws are constrained by bite-blocks appears to involve goal-oriented automatized motor

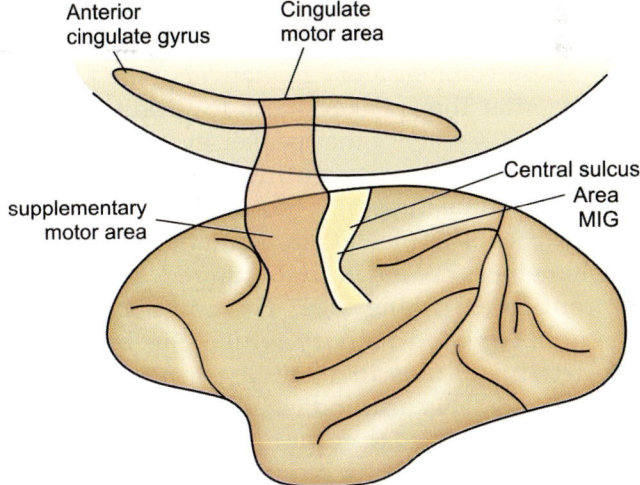

Fig. 9.4: Motor area

patterns. A bite block is a device that the speaker clenches between his molar teeth as he talks. The bite block can be made up of a small piece of wood or dental material with a string attached to it to prevent accidental swallowing of it. The movements of the speaker's jaws as well as the electrical activity of the muscles that close the mandible can be monitored by tapping the electrical activity externally or by electromyography. Four adult normal speakers were monitored during the production of repeated sequences of syllable [pi] under a normal condition. Bite blocks were inserted between the upper and lower molar teeth. The gap between the subject's central incisor teeth was 15 mm. The subjects compensated for the bite blocks by appropriate adjustments in their tongue and lip movements to produce the same acoustic signal [pi] under normal control conditions. Apart from these adjustments, the automatized muscle commands for the production of the syllable [pi] were also activated even though the subject knew that his jaws cannot move. The result of many bite block experiments suggest that the speakers have some sort of mental representation, of the supraventricular vocal tract (SVT) shape, that they normally use to produce a particular speech sound in a given phonetic context. Bite block induced deviations from the normal conditions result in the 'speakers' producing the appropriate SVT shape by means of compensating motor activities. These compensatory maneuvers take place within milliseconds after the speaker starts to produce the sound, excluding the possibility of the speaker monitoring the sound first auditorily and then executing compensatory muscle commands. This time interval is too short for the cross-modal auditory feedback. This mental representation of the SVT shape for each phoneme is not innate, but acquired, as the phonemes are different in each language. The compensating action involves goal-oriented activity which can be effected by different patterns of muscular activity, in the context which are goal-oriented. This also indicates that the movements of articulation are not related to gestures but it might be otherwise.

Thus, we can find that Dr Ramachandran's hypothesis goes very near to the concept that the word that is the outcome as a symbol for an object or action is not totally arbitrary but it forms a nonarbitrary relation between the visual sensation and the sound used for it for some dominating character between the two. He has however completely omitted the concept of a phoneme which by permutation and combination with other phonemes forms the word. It is mandatory therefore to explain how the first phoneme of the word forms the nonarbitrary relation between the dominant character of the profile of the object and the first phoneme of the word which will be the symbol for which it is used?

Other shortcoming in his concepts are mentioned above.

10 The Author's Hypothesis of Origin of Speech and Language

In this chapter the author has proposed his hypothesis of evolution of speech and language. He has already taken a review of various hypotheses proposed by many linguists and other scientists. He has also taken into consideration hypothesis proposed by Dr Ramachandran. It is the only hypothesis proposed by a neurophysician or perhaps to the best of my knowledge, from any other medical practitioner from any other discipline. We will also find out if there are some shortfalls in Dr Ramachandran's theory. Later the author has proposed his own hypothesis of origin of speech and language.

We have already seen the theory of origin of speech and language as proposed by Dr Ramachandran. In fact, rarely neurologists and laryngologists have taken interest in the origin of speech. Recently a branch of phonosurgery is evolving in the field of laryngology. This branch is concerned with voice and speech disorders. None of them, however, have thought of how this speech might have evolved 150,000 years BP? Some linguists, however, have published treatises on the evolution of speech. We have already taken a review of them in Chapter 7. Dr Ramachandran being a neurologist has gone deep in the role of functional participation of the brain in the production of speech. His stress is on the first word formation which forms a nonarbitrary relation with the visual profile of the object or the action which it stands for. This relation depends upon cross-modal-abstraction between the two in the presence of mirror neurons. The nonarbitrary relation between the visual sensation on one hand and the undulating (waveform) sound on the other. He says it happens by abstraction (adaptation) of finding the common denominator between the two, and grounded in some **abstract mental space.** My task is whether one can take up the concept further and explain the pairing of relation between the two on some logical explanation and physical relations, instead of taking resort to some abstract mental space. Also taking the relation between the two with some added characters to its logical end by uttering sequentially phonemes—a word.

Human vocal tract appeared half a million years ago. But the specific configuration competent for the production of speech had not evolved till the evolution of *Homo sapiens*. The right angle relation between the back end of the palate and the pharyngeal or the posterior 1/3 part of the tongue is essential for the utterance of certain phonetic units such as [i] or [u] and to a certain extent [a] (as in about). *Homo sapiens*, in short, were capable of uttering all the phonetic units or phonemes which included vowels and consonants. Our immediately previous species, namely Neanderthals were not able to do so. We have already seen that in detail in Chapter 5.

We have noted some evolutionary changes having taken place in the new species *Homo sapiens*.

Because of full biped position they had flexed skull and forward bending of forehead because of more growth of prefrontal part of the brain, resulting in increase in weight of prefrontal area. The jaw has receded back and the distance between the back border of hard palate and front margin of foramen magnum is reduced. Because of acquiring full biped position two things happened. One is the larynx was pulled down to its maximum extent and along with it, the posterior 1/3rd part of the tongue was pulled back and down facing backwards and forming the anterior wall of the laryngopharyngeal part of the pharynx now the cavity of mouth and pharynx were at right angle to each other. The remaining anterior 2/3rds part of the tongue in the mouth was thin and more agile. What was the net result of all these changes?

Think of a period 150 to 200 hundred thousands of years ago when *Homo sapiens* evolved. An individual member of the species might have just elevated his tongue up towards palate while vocalizing. He found out to his surprise, that the voice he produced this way, was different than an usual vocal call. Can we call this act of elevating the tongue as a variable? Further by elevating the tongue at various positions in relation to palate he could produce variety of different utterances. This was fitting in the inquisitiveness which is a character of modern human. These new sounds were latter given a symbol as vowels. Why vowels first we will see later? Up till now he had elevated the tongue but not touched the palate as he was not doing so, during the vocal calls (Figs 10.1a to c).

Some other time he might have accidentaly touched the palate with the tongue. Voicing took place at the same time. The sound produced this way, was clear, sharp and different than sounds he had produced during the prior vocalizations. This was the origin of consonants, naturally he must have tried this new ability by touching the tongue at various places of the palate. As we have already seen interdomain cognitivity or

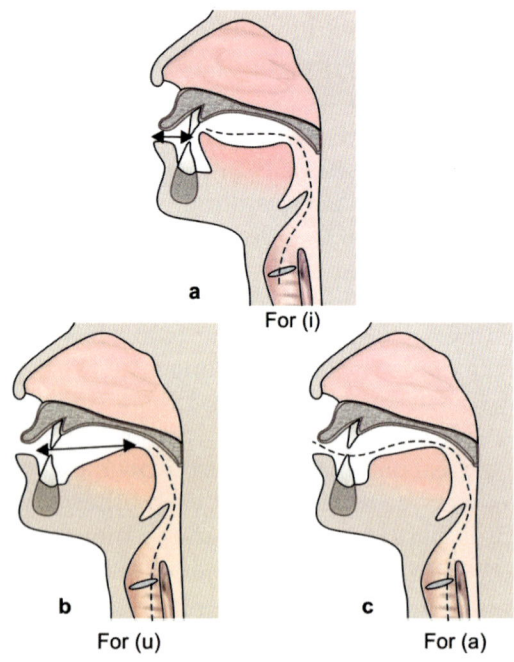

a
For (i)

b
For (u)

c
For (a)

Figs 10.1a to c: SVT; a. for (i); b. for (u); c. for (a)

'cognitive fluidity' had evolved in *Homo sapiens*. This level of cognitivity gave him ability to appreciate the difference in the quality of phonetic sounds produced by elevating the tongue and sometime touching it to the palate and the vocal calls he was using before for communication. Thus, the modern human had two more phonetic units in his repertoire. Combining the vowels and consonants must have followed automatically. Thus, with the combination of the two by permutation and combination process he was able to produce infinite number of phonetic units, **the words**.

This process of evolution and appearance of new sounds, their appreciation, all this did not took place in one day or in one generation. It must have taken, thousands of years and many generations. He had not still achieved full-fledged language. The evidences of cultural development in modern human species are available from 80,000 to 100,000 years onwards. This is a supporting evidence that till that time full-fledged language had not developed. Full transition from Hmmmm language of pre-human species had not taken to compositional language. The evidence of cultural development, and appearance of symbolic elements is a sign of development of modern or compositional language. Development of compositional language goes in tandem with more development of cognitive fluidity.

Earliest evidence for such developments, which we have already seen comes from Blombo's caves dating back to about 70,000 to 80,000 years BP and klassy river mouth region between 72,000 and 80,000 years back. The oldest fossil of *Homo sapiens* found, from out of Africa, are from Israel at Skhul and Qafzeh. They are dated around 100,000 years BP. This indicates that compositional language just had not evolved, in its todays full-fledged form. In tandem with the evolution of *Homo sapiens* around 150,000 years ago. The beginning of the development of words and language had taken place by the appearance of vowels, consonants and their meaningful combinations.

It is most natural thing to happen, that, the modern human must have thought of using the words formed by combining these units by permutations and combinations for different objects around and actions. Once skill for the production of vowels and consonants, the basic units and their combinations was achieved it was naturally a better option than the Hmmmm vocalization and gradually Hmmmm was replaced by compositional language.

In chapter three we have seen speech as defined by Nobel laureate Tinbergen as "Rapid stream of sounds (phonetic units) extended in time, we produce, by vocal apparatus." He had raised two questions. How are these rapid streams of sound produced? and who orders them? We add to it one more, who decides what word to be used for which object and action? That is the real origin of speech, and compositional language follows.

Modern man thus had vowels and consonants in his repertoire. He had also learnt to combine them in larger units. The art of combining small units into larger units was already there. As per Alison Wray the primates and pre-human species were segmenting their vocal calls and combining these to get new holistic units. If that is accepted the result will be a syllable. A phoneme has got no meaning. But a syllable and words formed from them have got a meaning. One or more syllables come together to form a word. Just as phoneme is the smallest unit of speech the word is the smallest unit of language.

When the modern man had syllables and words in his repertoire he must have thought of calling the objects around or the events taking place by the use of these words. The basic question was what combination of vowels and consonants or in short what word to be used for what object around? Is it completely arbitrary or is there any relation between the two? As the words for the same object are different in different languages of the world, linguists are firm on the acceptance, that the relation is arbitrary.

Or the process, was something like this, that one day the elderly members of the group sitting around the fire decided that let us call this object flying up in the sky a bird. And bird became a word for that object. Dr Ramachandran has said that this is absolutely a silly idea, of formation of first word. This cannot solve the idea of naming innumerable objects around. We have however learnt from Dr Ramachandran's hypothesis that there is some non-arbitrary correspondence between the visual shape of the object and the sound that might be its partner. This bias he says might be preexisting and hard-wired. But there cannot be hard-wired, preexisting bias, for every kind of object you come across or you are going to come across in future. However, one thing is certain that there must be some parallel between the two (or nonarbitrary relation), the object and the sound associated with it as its partner. He has also argued that it is arbitrary but becomes nonarbitrary grounded in a true resemblance of these two in a more abstract mental space. We want to see whether that nonarbitrariness: comes from some physical characteristics similar between the two and avoid the term "abstract mental space."

Dr Ramachandran in his hypothesis has never made a mention of phoneme. The words, however, are composed of phonemes and as Tinbergen has said speech consists of rapid streams of sounds extended in time, that we produce with the vocal apparatus. Utterance of stream of sound is a temporal event and they come one after the other. Dr Ramachandran in considering the nonarbitrary relation between the word and the object for which it stands for, has only thought of word and not about the phonemes which form a word. It is actually the first phoneme of the word which must form a nonarbitrary relation with the characters of the object then only the word will follow to form a relation parallel to the object. As we have already seen consonant which follows can be any one. Therefore in the discussion of our hypothesis we have mainly taken into consideration the first phoneme that must form a nonarbitrary relation between the sound and the object as a vowel. Sometimes however, the first phoneme can be a consonant, but the thought of having parallel relation of the first phoneme with the object or the action remains.

Let us look at it from a different angle. The object gives a visual sensation. The visual sensation coming from the object has got some characteristics. The object can be big or small, tall or dwarf, the action observed can have a speed fast or slow, etc.

Can there be some common denominator between the characters of the object and the characters of the sound that might be its partner? Yes, sound also has some characteristics. The sound is propagated in the form of waves in amedium.

(i) The sound wave has got a wavelength, (ii) sound has a frequency, (iii) sound has an intensity or loudness, (iv) the sound gets sometime distinctive features according to person or the object from which it originates. In case of human, the contents of the SVT the exact length of the SVT in each person, might have slightly different characters or dimensions. So also the fundamental frequency or fo might be a little different though within limits, which may produce these distinctive features in case of each individual.

In chapter three we have seen that the characters of the sound uttered during phonation such as wavelength, frequency, etc. depend upon the configuration of the SVT. Energy comes from the glottal opening in the form of fundamental frequency. Characters of sound wave are seen in Fig. 10.2.

The shape of the SVT can be changed by the articulatory organs, mainly by the tongue (Figs 10.3a to c). (Tongue positions in different vowels.) If the tongue touches or is elevated near the front tip of the hard palate then, it causes an obstruction to the flow of air or acts as a filter and divides the SVT in two parts. The part in front of the obstruction caused by the tongue

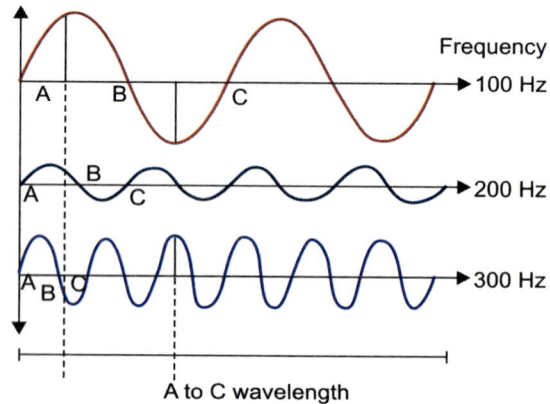

Fig. 10.2: Wavelength and frequency

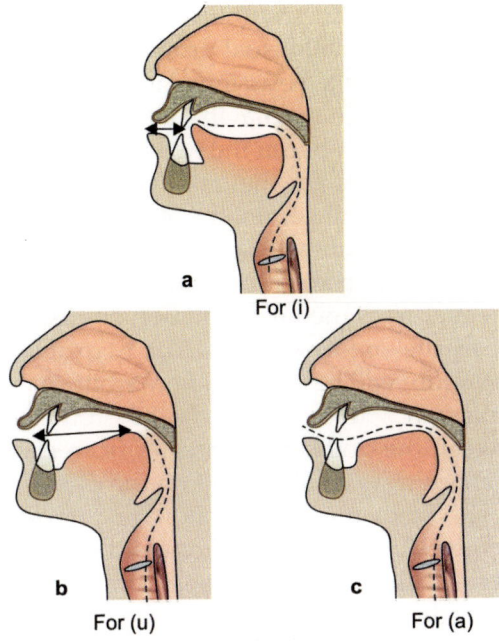

For (i)

For (u)   For (a)

Figs 10.3a to c: SVT with different tongue position

is smaller in length and the part behind is longer. When the tongue moves away the air is released and it has got different frequency and wavelength depending on the length of the part in front of the obstruction. (This we have already seen in chapter 3 but repeated for clarity.) This frequency is called formant frequency or F1. The wavelength of this sound is small as it comes from the small part of the SVT. The frequency will be high as frequency is inversely proportional to the wavelength of the sound wave. The sound uttered is called as vowel [i] as in heed]. [i] is a graphemic symbol we have designated for this sound. The symbol is arbitrary. Vowel [i] therefore is having a wavelength which is small and frequency is high, it is about 370 Hz. If the tongue on the contrary touches or is elevated near the most posterior (backward) part of the hard palate the situation will be exactly opposite. The part of the SVT in front of the obstruction will be long and the sound uttered under these conditions will be having a long wavelength and low frequency. We call this [u] [as in boot].

Now we have two things with us. An object and a sound. If the object is having a small visual profile the sound or the first vowel used for the word to be assigned to it will be coming from the small length of SVT and having a small wavelength. The common denominator between the two is smallness. The object is small, the vowel used for that is having small wavelength. There is one dominant denominator of the object, small or big and one dominant denominator of the sound wavelength small or large, which come from the adaptation of visual sensation and sound wave, cross modal adaptation or abstraction as it is also called, in the presence of mirror neurons. The sound produced therefore will be having a non-arbitrary relation with the object or action with the sound, based on the adaptation of the two between the dominant character from each. Thus, the relation is nonarbitrary and is based on the physical characters of the two. The other phonemes follow to complete the word used for the object. What is important is the characters of the first phoneme which must form a nonarbitrary relation with the object. The consonant used to complete the word can be anyone that comes to the mind that time, what is important is the vowel. Prominent vowel will be a front vowel for the smallness of the object, and a back vowel for the large object. Why importance is given to the vowel and the consonant used is accepted as of secondary importance, we will see later. Sometime the first phoneme can be a consonant as stated above, but the non-arbitrary relation between the object and the sound wave (wavelength) of it remains the same.

Most important question that arises here is, who decides or extracts the parallel characters between the two and forms the word?

Here we will recall our knowledge of IPL or inferior parietal lobule area in the brain (Fig. 10.4). We have already seen that IPL is a crossroad for three important sensations, visual sensation coming from the angular gyrus, auditory sensation coming from Wernicke's area, and joint and muscle tone sensation in parietal lobe. This area is very rich in mirror neurons. The Wernicke's area is the last station for the incoming auditory sensation. (Here the incoming sound wave, which is actually in the form of a parallel electric wave which is decoded or

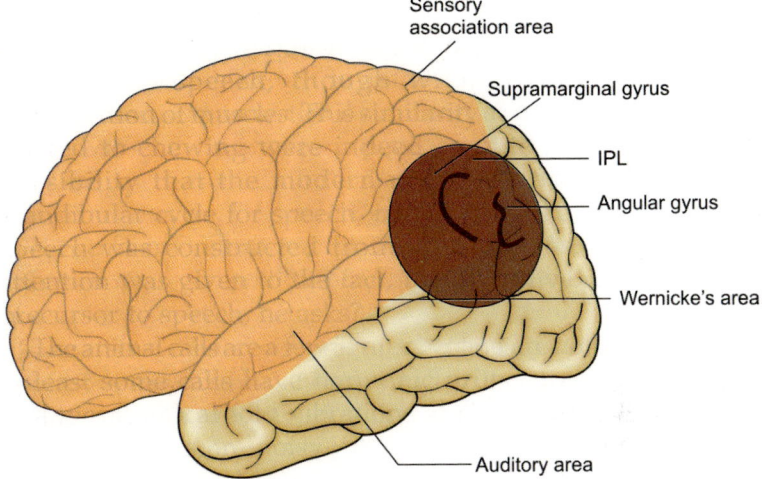

Fig. 10.4: Inferior parietal lobule

comprehended.) It is very natural therefore when one wants to speak, or for output, the sound wave is generated from this terminal area.

There are two sensations coming in, here the sensation from the visual profile of the object and the sound or auditory sensation. Say the object is very large like a tower. We can understand that the 'bigness' of the object does not make any sound, but the vowel which will be used to represent, it will have characteristics which are parallel to the big profile of the object which will be a vowel with long wavelength and coming from the long section of the SVT. This function of finding a common denominator between the two, is the function of the mirror neurons which are plenty in number in this area. Here there is abstraction or the cross-modal-translation between the two sensations, visual and auditory (synesthesia of Dr Ramachandran). This translation is executed in the presence of mirror neurons. Thus, there is nonarbitrary correspondence and cross activation between the brain maps for visual and auditory sensations—a built in translation between the two maps. The vowel used for the word for large object will be naturally a back vowel, say [u] and the sound wave having a long wavelength.

For the vowel [a] (as in again) the tongue lies down flat in the mouth cavity. Only the central part of the tongue is a little raised towards the center of the palate. According to some this elevation is not active but may be secondary to the constriction of the pharynx from both sides. This contraction can be as much as 5 mm. There is practically no division of SVT and whole of the SVT is active in the production of the vowel. The length of the SVT is about 17.5 cm. Therefore the sound produced will be having a formant frequency of 600 Hz.

The vowels produced by the tongue elevated towards palate in front of this, are called front vowels and vowels produced by tongue elevated towards palate behind this area, are called back vowels (Fig. 10.5).

Some inquisitive reader may ask, and rightly so, why only size is chosen as a dominant character for synesthetizing with auditory wave and its dominant character? There are some more important prominent characters of the object, such as color, shape, quality of surface and so on, why one of them was not chosen? The answer to this question lies in two factors: One is phylogeny and the other is related to physiology of vision.

Our vision has evolved many millions of years before. Animals on lower levels of evolution have got no neocortex. Highest level of their neural system is thalamus and basal nuclei. This is called reptile brain. They have also got eyes as external organ to receive visual stimulus. But vision does not

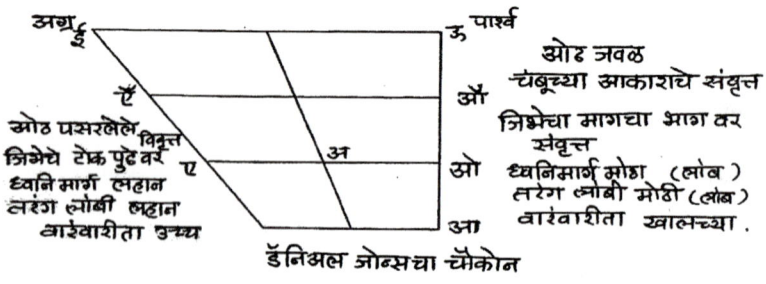

Fig. 10.5

occur in the eye. It occurs in the brain. Our auditory signal is comprehended only in area called Wernicke's area. But in case of vision it is not so. Carnivores and many of the herbivores have no color vision and have probably fewer than a dozen visual areas. But humans have as many as thirty visual areas. Why we higher primates have such a large number of visual areas, it seems they are all specialized for different aspects of vision such as color vision, seeing movement, seeing shapes, and so on. A good example of this is a tiny patch of cortex on both hemispheres that appears to be concerned with seeing movement, the area is called (MT) (*see* Fig. 10.3a). Let us first consider the pathways by which visual information enters the cortex. There is an old pathway present in animals, not having neocortex. This path starts in retina, relays through and ancient midbrain structure superior colliculus and then projects via pulvinar which is a part of reptile brain along with thalamus and other basal nuclei. In case of higher animals and primates when neocortex is developed this path extends to parietal lobe (Fig. 10.6).

After the development of finger grip with saddle joint of the thumb it was possible to catch a twig of tree judging the size and direction of the twig by vision, or a fruit judging its size and position. The link between the action and perception has become clear by discovery of new class of neurons, similar to mirror neurons called canonical neurons each canonical neuron fires during action similar to those mentioned above.

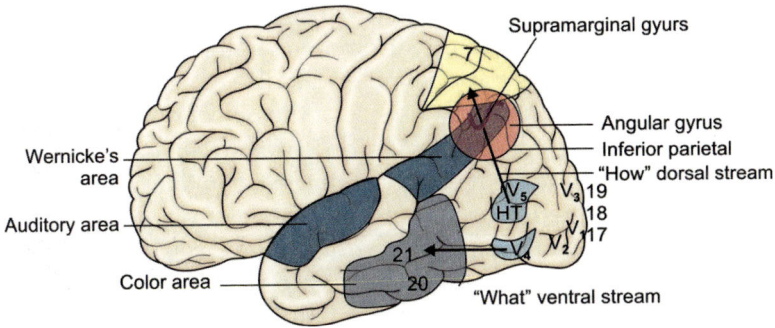

Fig. 10.6: Visual perception centers

In other words these neurons help in judging the abstract property of graspebility. In case of lower animals where there is no development of neocortex and therefore no frontal lobe there is no mention as I know, of discovery of such neurons though there must be some inherent property in their neural system for judging the size of the object for its gulpability. If a rat comes across an egg it can judge that it cannot be gulped and it breaks it and eats the contents, but if a snake comes across an egg it can judge its gulpability open its mouth and gulp. Visual appreciation of the size of the object is necessary for judging its gulpability. This is needed for survival of the species and therefore is a primary instinct. Even in higher animals when this pathway reaches to the parietal lobe it is concerned with spatial aspects of vision to know where an object is. Old pathway enables us to orient towards the object and exact orientation in the space is not possible without judging the size of the object in relation to the space. Thus, what I want to state that judging the size of the object has got prime phylogenic importance. What is physiological explanation.

The visual stimulus from retina travels to the posterior end of occipital lobe or area V_1. From their it splits into two streams. Pathway one or what is called as how stream (Fig. 10.6), reaches the parietal lobe. It overlaps the function of old path. It has got strong links to the motor system. It mediates much more sophisticated aspect of overall spatial layout of visual scene not only the object. This helps you to catch a dodge ball hurled at you. Most of these computations are unconscious and highly automated. You judge the size of the ball before its color and all other details. You are relying on the how stream for judging the size of the object hurled at you?

One more aspect of the vision by how stream is also important? For judging the place of the object in the visual space you have to judge its size. That can be done by automatically moving your vision round the borders of the object, if the object is large it will take a fraction of time in msec more (long wavelength) than if the object is small. Your reaction time is the same. If the object is small and hurled at more speed the reaction time interval between, the vision and the automated action taken will be naturally small (small wavelength). That may be a reason for choosing an auditory wave with long wavelength and low frequency and a wave with small wavelength and high frequency for cross-modal-abstraction or synesthesia between visual sensation and auditory wave matching it.

And what happens to the other stream starting from the area V_1. The path 2 or what it is called as what stream is concerned with details of other characters of the objects in the visual space. The stream projects from area V_2 to fusiform area in front and then to other parts. Here you evoke a penumbra of associated memories and facts about the object or the space in the visual field. Semantic retrieval process involves widespread activation of the temporal lobe with some "bottlenecks" (narrow paths) to include Wernicke's language area and the inferior parietal lobule, where angular gyrus represents the visual stimulation. This is involved in special human abilities such as naming, reading, writing, and arithmatic.

After the meaning is extracted here, the messages are relayed to the amygdala, a part of limbic system to evoke feelings or emotional expressions about what you are seeing. There is also a short cut to amygdala via superior temporal sulcus. This shortcut probably evolved to promote fast reactions to high value situations, whether innate or learned. Thus, visual sensation also reach the limbic system and adds to the generation of prosodic effects of language.

It can be appreciated thus that visual sensation and language are closely intertwined in each other and to study one we must have some information about the other, that is the reason this part is included here.

The tongue can move or be elevated at seven positions in relation to palate, three in front and three behind the central position for the vowel [a] (as in again). They are called quantal positions. The positions for the [i] extreme front and for [u] extreme back are called special quantal or steady state, as the tongue cannot take more front position than that taken for [i] and more back than that taken for vowel [u]. Tongue can go near the palate at more than these 7 positions but then our auditory apparatus will not be able to discriminate between the two adjacent phonemes. Another important point is, each quantal position has got an acoustic correlate in our auditory system including cochlea that for exact reception of the phoneme it should come from the corresponding quantal position.

The quantal positions for vowel [i] and vowel [u] are fixed as tongue cannot go either more in front of [i] or more backward than position for [u]. The other quantal positions are decided by in relation with the distance between these two. The utterance of the vowels in between these two positions and there perception may not be very accurately fixed but may have some variation.

In our above quoted example, therefore, the vowel chosen for the large object will be vowel [u] but which back vowel for the big or which front vowel for small objects, who decides that we will see later? The auditory sensation which is parallel to the profile of the object and which itself has got the information of the wavelength of the sound matching the pronunciation of the phoneme is carried to the Broca's area via 'arcuate fibers' from Wernicke's area (Fig. 10.7).

Speech in primates and pre-human species is accompanied by manual and other gesture (Fig. 10.8). Here a few words

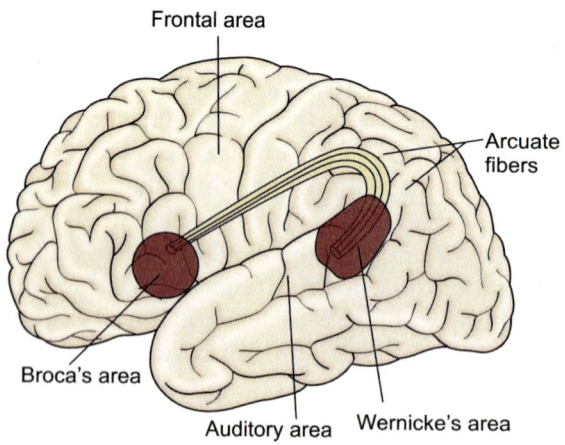

Fig. 10.7: Arcuate fibers

about gestures may help our further discussion. Gestures may be integral to human expression. Researchers are unanimous, on the point that out of many gestures, pointing was unique to human. Can it be because pointing is a function of supramarginal gyrus and it is only present in modern human? It is formed by the splitting of excessive growth of IPL, about 7 times than that in apes. Gestures come in variety of ways. Action gestures, abstract pointing, metaphorical gestures, symbolic reference to space or size (*see* Fig. 10.6), events, people, etc. Gestures amplify the meaning, sometimes communicate information that is not explicitly stated in verbal message. With gestures, listeners can even guess the meaning without sound. Researchers consider it as paralinguistic phenomenon, sometimes emphasizing the meaning. Gesture does not simply precede or accompany the language but it is fundamentally tied to it and act as a preliminary scaffolding of language even before a child has spoken a word (Fig.10.8).

It takes a while for gestures to take characteristic forms of a specific language. When it does people change their gestures depending on the syntax of the language they are now speaking. It is impossible to think of evolution of modern language without also considering the evolution of gestures. Precisely how speech and gestures may have interacted when we split from our common ancestors is still debated.

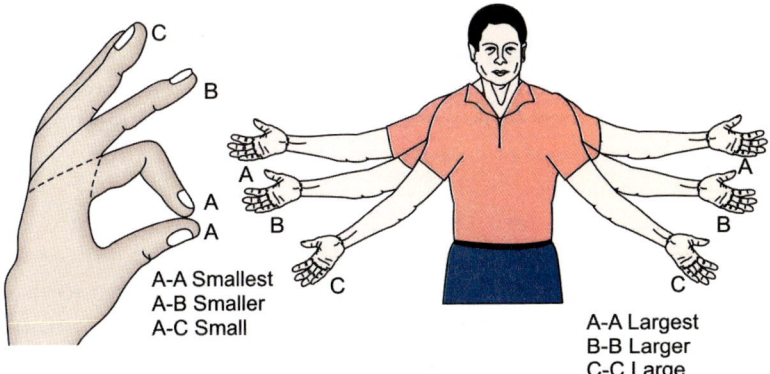

Fig. 10.8: Hand gestures showing size of object

One finding in case of apes is very significant. Apes have no speech or language. Their communication is mainly via vocalization and lot of gestures. Their Brodmann's area 44 on the surface of the frontal lobe representing the area for motor action related to gestures is very wide. The gestures accompany the vocal calls and they are also executed in response to visual sensation coming in from the space around, say as warning of predators, etc. The same area 44 in future acts as Broca's area in *Homo sapiens*.But the area is very much reduced specially, as much information can be conveyed by speech in case of humans. As more neurons are added for more actions, the region of area 44 in front of the area for action potentials for the movements of jaw, tongue, lips and mouth on precentral gyrusis markedly elevated and is now known as Broca's area.

There are already motor maps for manual gestures too, in Broca's area. These gestures are as a response to visual sensation coming from the profile of the object (*see* above). Broca's area is also rich in mirror neurons. In the presence of mirror neurons there is cross-modal-abstraction between the visual sensation coming from the exact profile of the object and the motor action stimulus for manual gestures exactly parallel to the visual sensation coming from the object. We have already seen in Leiberman's theory (Chapter 7), that in apes in presence of mirror neurons, the input of visual sensation can adapt and link to auditory outputs. Thus, the visual sensation of gestures can act as input for visual sensation which easily links with the auditory sensation or auditory wave. This takes place in area 44 of Brodmann future Broca's area—area 44 is very wide in apes thus the place for function is presumably already there in our ancestors brain who lived 20 million years ago (Fig. 10.8).

If the object is very large the hands are spread very wide apart to show a movement parallel to its size, if the object is very small tiny the manual sign also will be matching to it to show exact smallness of the object very small by the thumb and index finger touching each other and so on. The cross-modal-abstraction here will also change the auditory sensation matching to the visual profile according to manual gestures. The characteristics of the manual gestures accompanying the visual profile of the object is the cross-modal-abstraction between sensory and motor maps and the auditory sensation translation, matching to motor maps of gestures in Broca's area, to exact wavelength of the auditory wave is between the motor to sensory maps (Fig. 10.9).

Thus, here is ultimately translation of visual sensation maps of object to the exact sensory maps of the auditory wave have taken place. Sensory to motor and motor to sensory. Auditory sensation then choose the exact wavelength of the sound wave matching the vowel to be uttered. The manual gestures per se do not accompany the speech every time. This is due to the inhibitory effect of frontal lobe which suppresses the actual act of expression when it is not appropriate to the occasion. The motor activity for the gestures however do take place in Broca's area in the form of electrical stimulation without actual execution. These are dummy movements. We have already seen about this phenomenon in our discussion about functions of mirror neurons in detail. Thus, when the auditory wave reaches the Broca's area there is a built in translation between the auditory maps and the visual maps of the profile of the object on one hand and the motor maps of the gestures on the other. There is cross-modal-abstraction and the auditory maps are

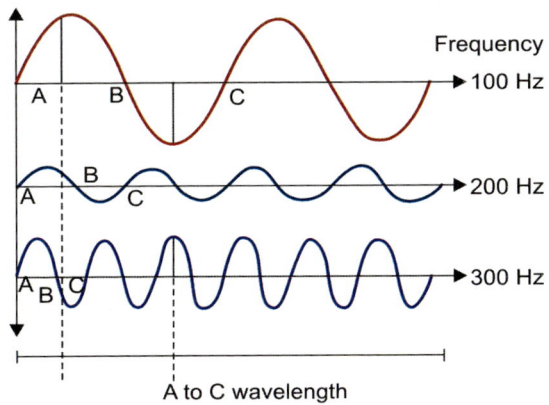

A to C wavelength

Fig. 10.9: Wavelength

now parallel to the motor maps of gestures and to the exact size of the object. The vowel to be chosen is having now exact parallel to the auditory map and an exact vowel having the information regarding the wavelength of the formant frequency of the chosen vowel. This is how the exact vowel having parallel wavelength is chosen? The gestures show exactly how big or how small is the object? The auditory maps are now equipped with the information of exact bigness or smallness of the object. These auditory sensations now stimulate the Broca's area for articulation of that phoneme. The message from the Broca's area, as we have already seen, is 'goal-oriented' and not to the individual muscles required for the utterance of that phoneme. Thus, the auditory sensation reaching and stimulating the motor maps in the Broca's area which are goal oriented for utterance of designated vowel. This goal-oriented stimulus is equipped with the information of the movements of the articulatory system including the movements of the tongue which ultimately decides which phoneme is to be uttered.

Is that all in utterance of a phoneme? No, the speech has got one more very important characteristic. That is emotional expressions. The center for emotions is in limbic system which includes amygdala, anterior cingulate gyrus, thalamus, hypothalamus and other basal nuclei (Fig.10.10a).

How are the characteristics for emotional expression are added to the speech? These additional characteristics including emotional expressions are called suprasegmental or prosodic features. They are also called paralinguistic effects. How the sound wave or the stimulus reaching the articulatory system changes due to prosodic effects? Does it changes the wavelength? Does it changes the frequency? No. Otherwise the phoneme itself will change as which phoneme is to be uttered is dependent on the wavelength of the sound wave.

Prosodic Features

Variation of stress, intonation and duration or lengthening of the phoneme convey the linguistic, paralinguistic and emotional information. Variation in stress (Daniel Jones uses the term prominence), have systematic linguistic function for many languages.

It can change acoustic effect (Fig. 10.10a). Tone plays important linguistic role in many languages such as English, Swedish, Thai, etc. In Chinese, two words may differ in meaning solely on phonetic basis with respect to the tone pattern of fundamental frequency. In these tone-based languages the speaker continuously maneuvers changes in patterns of fundamental frequency contours to produce the necessary effect. The pattern of fundamental frequency plays a role in signalling the end of the sentence in most of the human languages.

We have already seen that the fundamental frequency is related to size and shape of the larynx and length of the vocal cord in male and female. It can also change with the change in tension in the cords. Thus, these slight changes can be produced by change in tension and length of the cord and length of SVT. The changes in the cord can be achieved with the help of internal laryngeal musculature and length of the SVT can be changed by other maneuvers such as protruding the lips and moving the larynx up or down as necessary. These maneuvers can change the fundamental frequency and produce the required effect. Protruding the lips lengthen the length of SVT, bending the neck down reduces the length of SVT and reduces

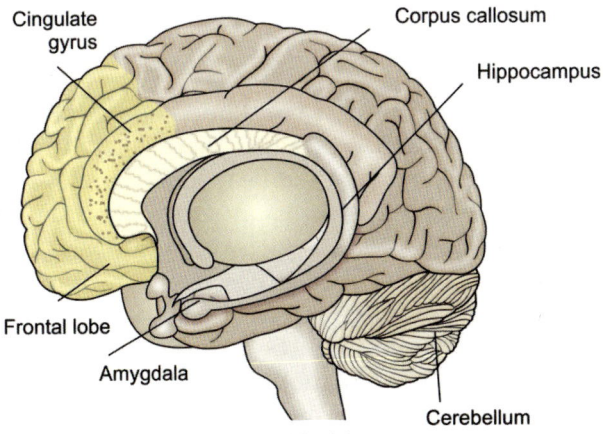

Fig. 10.10a: Limbic systems

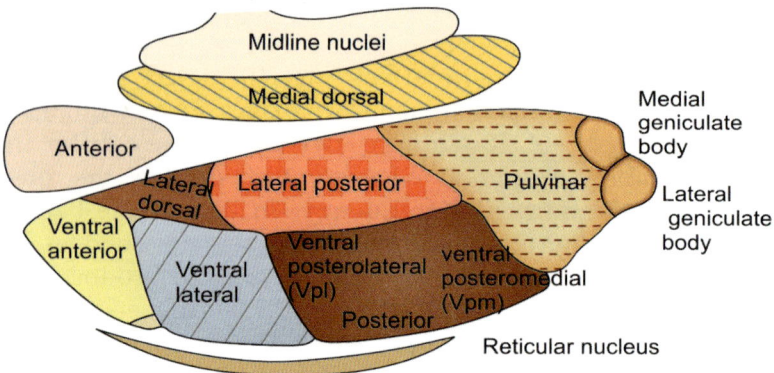

Fig. 10.10b: Basal nuclei

the tension of the cords. Extending the neck up and back increase the tension of the cords.

How change in fo or fundamental frqueny of voice energy from the glottis can produce the prosodic effect we have already seen in Chapter 3. The filter or obstruction to voice energy divides the SVT in two parts. The formant frequency for a phoneme arises from the part of the SVT in front of obstruction. After the formant frequency is released all other frequencies coming from the back part are dissolved. However, the fo or fundamental frequency from the glottis continues, and it mixes with the formant frequency which has already produced a phoneme and when these two frequencies mix there are peaks and troughs formed (beats).

The position of the beats is fixed and can be found out by mathematical calculation. But when there is slight change in fo it changes the positions of the beats and therefore changes the qualitative character of formant frequency this adds to prosodic aspects (Fig. 10.11).

If you ask someone to read a passage from which all the orthographic punctuations such as full stop, comma, etc. are removed, the experience will be an unintelligible speech. Therefore orthographic punctuations are included in prosodic features and are important. Thus, these symbols are essential. Question marks, exclamations optionally replace some special

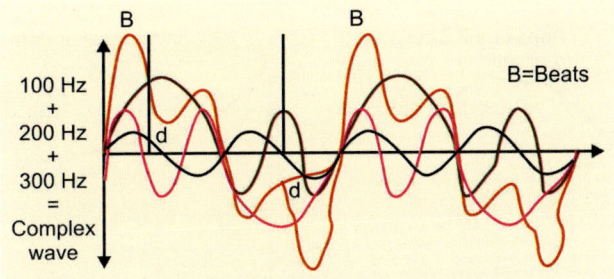

Fig. 10.11: Beats

words in some languages. Without these orthographic punctuations and symbols the sentence will be like a train of words and will be difficult to decode. The phonetic features that speakers make use of, to segment the train of words into a meaningful sentence are executed with the help of the use of breath group or required expiratory pressure. Without these breath groups and prosodic features the language will be reduced to one word utterances each of which have a fixed immutable meaning.

Language is not a code in which a particular signal (words and punctuations) have fixed meanings. Language has got a potential of translating new unanticipated information from the same words. Breath group or change in breathing pressures are the basic primitive phonetic features that serve as an organizing principle for these factors. Breath group or breathing pressure falls at the end of inspiratory cycle (Fig. 10.12).

The variation in breathing pressure can change the characteristic of the phoneme by adding stress, lengthening the duration, etc. though the basic sentence remains the same. For example look at the following conversation between father and the son.

'Yesterday you went for a movie' is a statement. If stress is given on the word 'movie' that becomes a question and if stress is, on all the words with rising pressure in the end will make it an exclamation, meaning, I did not like it.

Every one has probably seen the ad of a car. The car suggest any car but lengthening the word caaaaaar makes it a luxury car.

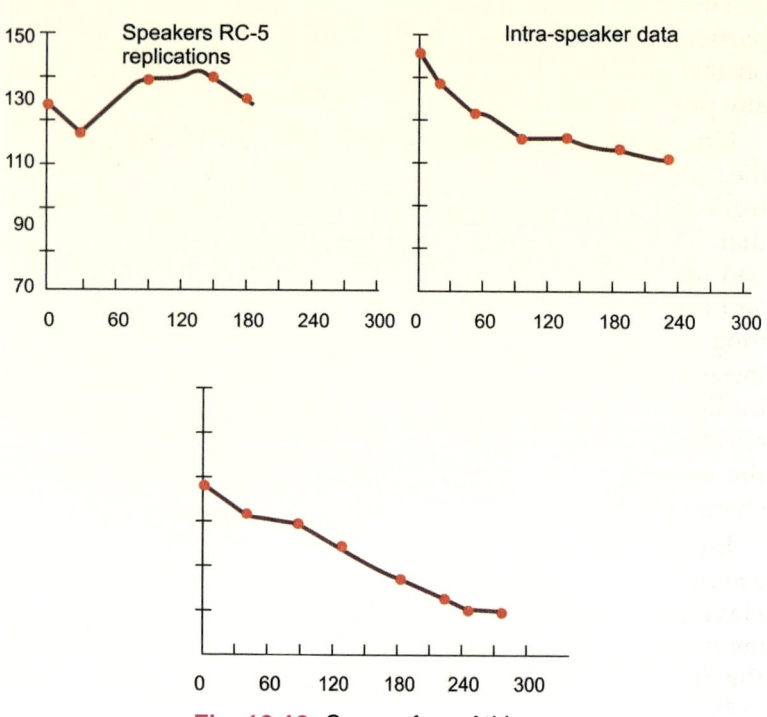

Fig. 10.12: Source from Atkinson

Note the duration in time in msec for each syllable. Duration of the last syllable is lengthened and the intensity is lowered, thus it furnishes a salient cue for the end of the sentence. Fundamental frequency also falls with increase in duration. In English, the question requiring answers in terms of yes or no, the yes–no are produced in a positive breath group.

Equipped with all these characters a formant frequency matching a phoneme which is matching the visual profile of the object, the gestural motor maps adding the extra information by suitably changing the formant frequency of the phoneme to be uttered and breathing pressure difference with slightly changing the fo or fundamental frequency and duration adding emotional and prosodic features to the phonemes and word to be uttered make a perfect meaningful speech. This is how the *Homo sapiens* sometime after 150,000 to 200,000 thousands of years BP uttered the first word.

Thus, far we have discussed the logical basis for using a particular phoneme for a particular object or action depending on the common denominator between the two. But is there any practical proof of this hypothesis? Yes, let us see what it is?

Otto Jespersen in 1920s (quoted by Berlin 2005) noted that the sound [i] comes to be easily associated with small and sound from [u, o, a] with bigger objects. He also gave a reason for that. The sound [i] is made by pushing the tongue forward and upward to make the smallest cavity between the tongue and the lips, while the [u, o, a] sounds result from a lowered tongue which creates a larger mouth cavity. He thought that these sounds are the physical gestures made by the tongue and the lips which mimic the size of the object being named. Though his observation was partially correct he was unable to connect the relation between the two depending on their physical characters. His claim was a little more than a hunch.

Edward Sapir is a distinguished linguist of 1920s undertook a simple test. He made up two nonsense words Mil and Mal. Having two different vowels, a front [i] (as in heed) and [a] (as in art) in the two words. He told his subjects these were the names of two tables and asked them, which word indicated the large table and which the small. He found as predicted that almost all had chosen Mal for the large and Mil for the small.

Could this same association of sound of vowels and object size have played a role in naming the animals by tribes in the modern human world.

Brent Berlin Ethnobiologist from Georgia university examined the names that the Huambisa community in Peru and Malay of Malaysia, used for fish. He chose fish so as to avoid any onomatopoeic influence over the generation of the word. In both the languages he found a very significant association between the size of the fish and the types of vowels used in their names. Those that are small in size are likely to have names using the vowel [i] for small fish, while those that are relatively large are more likely to use the back vowels [e, a, o, u] Huambisa and Malay live on the opposite side of the planet, it is very unlikely therefore that such similarities in vowel use, could have originated from a shared ancestral language. This is a manifestation of sound shape translation.

The same association was found among the bird names used by Huambisa and three other entirely unrelated languages, one of them was Tzeltals of Mexico.

Berlin also investigated, words used for Tapir a large slow moving animal and for squirrel which is quick and small, in nineteen South American languages. In fourteen of them the names for Tapir involved [a, o, u] or back vowels, while [i] was used for squirrel names. The vowels [a, o, u] were used as they represented sound translation for both size and speed. Tapir has got big size and slow speed therefore back vowels having long wavelength and low frequency were used and for squirrel which is small and fast in speed, the front vowel [i] having small wavelength and high frequency was used. This can be said as manifestation of sound-shape and speed translation.

The bird names of Huambisa tend to have relatively large number of segments of acoustically high frequency (front vowels) which appear to denote quick and rapid motion or what he calls birdiness, in contrast fish names have low frequency segments (back vowels) which have a connotation of smooth continuous flow, fishness.

Berlin further tested whether English speaking students would be able to distinguish between Huambisa bird and fish names. He took 16 pairs of words one of the name of a bird, the other that of a fish. One pair amongst them was chunchuikit and mauts, another was iyachi and apop. He asked 600 students to guess which was which. Their guesses were correct at a significantly high rate than would have been expected by chance alone. 98% rightly thought that chunchuikit was a bird name and mauts that of a fish.

These practical examples prove our point that there is not just any arbitrary relation between a sound and word used for an object or event but there is always some common denominator between the physical characters of both. The sound chosen is a manifestation of, sound-size/shape and speed translation based on this common denominator.

All the examples chosen above are mostly from tribal communities as their language is changed very little over the period.

Why Did the Vowels Evolved First?

What phoneme was uttered first? The answer is vowel. Why? Let us find out, whether we can justify. We have seen about the utterance of vowels matching to the visual profile of the object. We also have noted why consonants are of secondary importance? We have seen in Chapter 3, why vowels are more important in speech and not consonants? Now we have to find out an answer to our statement that the vowels were uttered first from the evolutionary point of view.

We use front vowels for small objects and fast movements. [i] (as in ease). We use back vowels for large objects and slow movements. [u] (as in boot) is the back vowel used for slowest movement. Other front and back vowels are used for the similar concept, varying only in degree respectively. Why we have chosen [i] and [u] only in our discussion? Roman Jacobson has shown that these are the vowels which come very often in any language. If we consider vowels only, vowel [i] comes 91.5% times [u] 88% times and [a] comes about 83% times. Another reason is these two vowels are special quantal as we have already seen. That is the reason why the utterance of these two vowels is accurate?

Gordon Peterson undertook one experiment. He made a list of words having same consonant but different vowels such as hood, heed, head, etc. He asked 76 volunteers both males and females to utter them, recorded and mixed them randomly so that it was not possible to know which word was uttered by whom. He then asked ten thousand volunteers to listen the words and repeat. He found out that only two mistakes were done in listening the vowel [i], five for vowel [u] and more than 500 for all other vowels. This proves why the vowels are more important and why we have taken [i] and [u] only for our discussion.

Let us find out now why we say that *Homo sapiens* must have uttered vowels first? The reason for that lies in phylogeny.

Primates and pre-human species were giving their vocal calls by keeping the mouth open and tongue lying flat on the floor of the mouth. Variation in calls was effected by varying the expiratory control and the fundamental frequency. In human speech however there is opening and closing of the mouth. The tongue is more agile and can do more movements. We have seen that sometime the human might have elevated the tongue towards the palate and gave a call. He must have found that the phoneme he produced was different, it was but natural for him to try with varying the position of the tongue in respect of palate and then found out that he could vary the sound he thus produced.

The important variation however was the tongue was raised but did not touch the palate, there was a little gap between the two. Keeping a gap between the two was the remnant of pre-human action of vocalization and had phylogenic reference. The vocal calls in pre-human species were given while keeping the mouth open, they had no open and close cyclicity just as in modern human or *Homo sapiens*. As there was a gap between the tongue and the palate some air must be skipping out which gave the vowel two characters. It had more duration and more energy in utterance. The vowel therefore gives energy to the speech. The duration of a vowel is about 200 msec to 300 msec. Phylogeny or Earnst Haeckel's principle of ontogenic recapitulation was thus responsible for the first utterance of vowels by the *Homo sapiens*.

The vocal calls of primates and pre-human species were originated in limbic system and from the right brain. In human the vocal speech involves the movements of tongue, jaw, palate and other articulatory organs. These organs were used more than half a million years for eating, chewing and drinking and therefore there were already neural circuits for their movements. There was splitting of these circuits, as it often happens in evolution and new functions were taken up by new circuits and old functions also continued (exaptation). The old movements of eating, chewing, etc. were executed by the left brain, the movements of articulatory organs for speech therefore also shifted to the left side of the brain, emotional contribution however remained with limbic system on right side and was added by thalamocortical connections. This is a process which is called exaptation.

Consonants

We have discussed at length how the first phoneme, usually a vowel is selected and uttered. Speech is sequential uttering of sounds, but word is not formed only by vowels there are other categories of phonemes, namely consonants, which are also important. The sequence of phonemes in the word might be formed by permutation and combination of vowels and consonants. It might be VCV (vowel-consonant-vowel) or VCCV, or CVC and so on. Consonant thus is an important part of the word formation, and in turn, of language.

In any language the number of vowels is always less than that of consonants. In English, e.g. there are only five [a, e, i, o, u] primary vowels. Their allophones may increase the number. However, consonants are always more in number. The phonemes are uttered based on Johannes. Muller's (1848) source filter theory. The filter is either lips or the tongue. The tongue touches the palate at various positions for the utterance of different consonants. These are known as quantal positions. There are 7 positions including one at the front and one at the back end of the palate. The tongue causes complete occlusion of the SVT.

The consonant is released when the lips are parted or the tongue shifts its position. These consonants are called plosives or those which are uttered by burst of air after the release of the filter. In such a case there will be only seven consonants uttered depending on which position the tongue reaches. These quantal positions are fixed because there is acoustic correlate with these positions in the cochlea. Further the seven positions are linked with corresponding neural property detectors in the auditory area in temporal lobe. Our auditory system is evolved in that way.

However, these quantal positions are vary ingeniously utilized for the production of five different consonants each, from tongue touching the same quantal position.

In our vocabulary some consonants are voiced and some are voiceless. This difference depends upon voice onset time (VOT). What is VOT? As we know when a phoneme is uttered two actions take place. One is shifting of tongue from the quantal position and at the same time burst of phonetic sound from the glottal opening by parting of the vocal cords, which are already in closed or phonation ready position. The cords are pushed away at the same time by the air pressure built up below the cords. These two things, shifting of tongue and burst of air take place almost simultaneously. Consonants uttered in this process are called voiced consonants as the burst is accompanied with phonetary or vibratory sound. They are [b] a labial and [d] a dental respectively, when the lips or the tongue is acting as a filter. In case of some consonants, however, the process takes place in a bit different sequence. When the filter either lips or the tongue moves away the vocal cords have not reached their closed or neutral position. The vocal cords are still slightly open at the instance when the filter moves away. The sound that initially occurs is not voiced since phonation cannot occur when the cords are open. The open position of the cords however allows a relatively large air flow and the initial sound is generated by the air turbulence that occurs as the filter is abruptly opened. And the cords are not still closed. Initial airflow or the burst, therefore can occur when a phoneme is produced before the cords are closed. The speaker starts closing his vocal cords after the phoneme is released. The phoneme might be [p] when lips act as filter or [d] when tongue near dentogingival margin is acting as filter. The difference between [b] and [p] or [t] and [d] thus depends upon the delay in the start of phonation in relation to removal of filter. This difference is depending on phonation onset, VOT. The difference in timing of the removal of the filter, either parting of lips or shifting of tongue to next position, must be more than 25 to 30 msec. For the two consonants one voiced and another voiceless to be detected as separate consonants, i.e. [p] from [b] or [t] from [d]. If the difference is less than that the two will be identified as one, the voiced only.

Thus, we have got now two sets of consonants, from each of the filters lips and tongue, at the roots of teeth. [p] and [f] as in fur both voiceless, and [b] and [t] as in tea, both voiced. And [f] as in fur and [dh] as in dhow (a small ship) are both voiceless. Now we have to solve the problem how to differentiate between the two voiced and two voiceless consonants? At the end of pronunciation of two voiceless consonants [f] and [dh] air flows with friction along the tongue which is moving to another position thus they are fricatives or aspirants and the two others [b] in the end while the two other voiced are nonaspirants. Thus, we have got four different consonants from each filter positions.

T	D
Voice less nonaspirant	Voiced nonaspirant
TH	DH
Voic less aspirant	Voiced aspirant

Experiments have proved that, infants one-month-old, are also sensitive to this 25 msec time distinction. It is not possible that the infants of that age could have learnt to respond to speech stimuli in this manner, therefore the 25 msec time distinction reflect an innately determined constraint of the auditory system of *Homo sapiens*.

Studies have shown that even 'chinchillas' (a small animal like rat with grey fur) respond the same way to this 25 msec threshold. These results show that these capacities were developed in mammalian auditory system as a result of Darwinian process of gradual evolution and *Homo sapiens* have taken advantage of these capacities for their speech and language development.

Apart from these four different consonants produced from the same filter position, one more consonant is produced from the same position and that is the nasal sound [m] labial or [n] dental. Nasal sounds in English are similar to stop consonants. However, in contrast to oral stop consonants the velum is open and there is gap in the postnasal cavity through which air can escape. The sound is propagated through both the nose and mouth. The nasal consonant has got an additional property associated with the presence of a murmur in the vicinity of the consonant release. The air passing through the nose gets resonance from the hollow cavities of the paranasal sinuses. The acoustic property of the murmur therefore a broad resonant peak around 250 Hz. And with other resonant peaks of low amplitude above 700 Hz. The nasal murmur must occur while the closure interval of the SVT and must continue just before the release of occlusion by filter, otherwise the consonant will not be heard as nasal. There is presence of anti-formants or zeros which selectively absorb acoustic energy apart from formant frequency.

The nasal murmur is not the same for all the nasal consonants. Under these circumstances the onset of phoneme specific formant frequency is sufficient for the perception of nasal consonant irrespective for the place of articulation.

Thus, we have discussed in detail, how only seven quantal positions of occlusion, can produce five different consonants having different acoustic characters.

Not all consonants are plosives or produced by occlusion and burst of air. Some are released by voicing with air passing with friction against organs in SVT mainly the tongue or gliding by the side of the tongue those are the details of articulation and part of structural linguistic and therefore are beyond the scope of this treatise which is mainly concerned with the evolution of speech.

How Word is Formed?

Tinbergen has defined speech as a "Rapid stream of sounds extended in time that we produce with vocal apparatus." The definition indicates that the stream of sounds is a temporal event. That is one sound is followed by the other and each have independent existence. Linguists had thought the sounds of speech were, similar to beads on a string, each independent coming one after the other. In such a case we would have heard (Fig. 10.13) one sound coming after the other and not a syllable or a word.

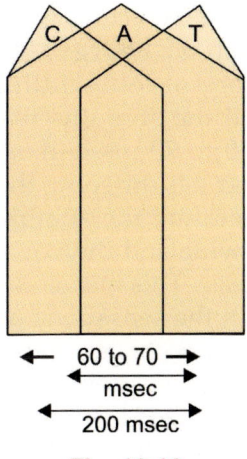

Fig. 10.13

Fortunately it is not so. We have already gone into the details of anatomy and physiology of hearing in Chapter 3. Let us take a review of relevant facts that will give us factual knowledge of how a word is formed. As we can imagine word or syllable is cluster of phonemes having meaning.

Each phoneme has got a characteristic formant frequency which stimulates an organ of Corti, which is tuned to that Formant F. The frequency of the organ of Corti is dependent on the length of the bundle of fibers of the basal membrane on which it is situated. The length is factual and tension depends upon the supporting cells placed on that fiber bundle. Here in organ of Corti the sound wave is translated into an electrical wave. We have already seen about it. If only this wave reaches the auditory center we will receive the sensation of that particular phoneme but it is not that simple. Here we have to recall our knowledge of 'perception of vision.' Even if the object is removed from the visual field we perceive the same image for sometime. Does it happen in the same way in case of sound? Yes, but a little differently. The organs of Corti are situated so close to each other the impedance will certainly carry the vibration to neighboring organs. This is known as band spread. Width of the band will depend upon the strength of the original stimulus. The consonants give clarity to the speech but not much of energy. The band spread of consonants is therefore of less width and we will get an auditory sensation till the stimulus reaches the end of the band. This is about 60 to 70 msec in case of consonants. In case of vowels however it is different, the vowels give energy to the speech and therefore the stimulus is strong and the band spread is more wide and the vowel stimulus we listen for about 200 to 250 msec. This means sensation of vowel will cover and mingle with the sensations of consonants coming before and after it. Thus, a cluster of two or more phonemes will be heard as a one word. Word can be of one syllable or more. After each word there is a small gap of few msec, which is not clearly shown in speech, but is represented in writing by breaking each word from the other. Take an example of a word cat. The central vowel is perceived for more duration and it covers consonants and it forms a cluster of phonemes which forms a meaningful word cat (Fig. 10.13). This is how a word is formed? There is slight change in the phonetic configuration of the phoneme when it comes in a word, known as 'allophone' but it does not change the phoneme. This is one of the cues how we understand where a word starts and where it ends in a stream of sounds of speech?

The line between the visual perception and visual imagination became increasingly blurred in human evolution. A monkey or a dog probably enjoys some rudimentary form of visual imagery only human can create symbolic visual tokens and juggle them around in the mind's eye to try out novel combinations such as babies sprouting wings (angels) or beings that are half horse-half human. Only human can juggle the symbolic tokens such as alphabets for, uttered sound waves, which are arbitrary and try out novel juxtapositions. Such off-line symbol juggling may in turn be a requirement for another unique human trait, word formation as basic art for the origin of speech and language.

There is one more cue how a child learns where a word starts and where it ends, and to which grammatical category it belongs to, in a continuous speech signal? There exist multiple sources of information about it that allow an infant to take probabilistic approach, which in turn, can produce good guesses (Seidenberg 1997). Here we can recall the work of Jenny Saffron. Jenny Saffran a developmental psychologist in the university of Wisconsin-Madison, thinks, infants are able to identify statistical regularities in continuous speech they hear as parents talk. When infants hear 'parents talk' or IDS (infant directed speech) they do not hear just a stream of sound but also 'clusters' of phonemes forming syllables and identify them. She says infants are born statisticians. When infants are familiarized with a stream of sounds in which some syllable sequences are always clustered together. After elimination of prosodic cues, infant mind is busy extracting the statistical regularities introduced by the recurring cluster patterns. If that is correct then this ability must relate to how we get the most general acceptance of the words.

What Word First?

What word human might have uttered first? Was it a verb, a noun, a name of an object?

The primary symbols of communication are a cluster of phonemes grouped in the form of a syllable. Syllables form words. Language in its infancy was not concerned with naming the things, or individuals but with relationship maintained by action between the members of the group. Things and their relations were considered as a single individual whole. This might be polysyllabic, but its fraction, would posses no significance when pronounced separately, or as Jespersen (1922) put it primitive linguistic units must have been much more complicated in point of meaning, as well as in point of sounds, than with which we are familiar today. The evolution of language shows a progressive tendency from inseparable irregular conglomeration to freely and regularly combinable short elements. Even today, some tribal languages cannot describe father except as, my father or thy father. The word without pronoun is meaningless.

Diamond's view is that the evolution of speech is to be sought by discovering what kind of word is most common in the earliest form of any language and in the oldest language. Every complete communication includes a 'verb' and that the proportion of verbs to other part of speech decreases as a particular language develops. The second person singular, the request for action is the simple form of the verb without affix and without rules of syntax. He has investigated many semitic languages which he says have changed, far less in prehistoric times than any other language known to us. After studying the Bantu tongues which have spread over Southern Africa within the last two millennia he has reached to the conclusion that the earliest words were 'verbs' of action and earliest forms the second person singular. When he counted words in these languages he found the number of verbs exceed than any other category of words even today. If we come across an individual from a remote tribe he automatically restricts his talk to more verbs of action than any other category of grammatical units.

Application to Today's Language

We have seen how speech and language evolved 150 thousands of years onwards since the evolution of *Homo sapiens*? Our theory has most logical foundation. One will naturally ask whether it can be applied to today's language? The compositional language since ancient times has evolved with such incredible rapidity; sometimes in just two hundred years. A young person of today might not be able to communicate with his great, great, great-grandfather. The languages have changed countless number of times without reckoning. This can be as a result of migration, environmental changes, cultural evolution and many other reasons which cannot be identified. The great epic of Hindus in India, Bhagavad Gita was originally written in Sanskrit, which was a dead language of communication in 13th century, therefore it was translated in prakrit the language of communication then. Even the prakrit has changed several times since then and the epic had to be rewritten intodays Indian languages. Language changes completely in about 1000 years but language is never dead. Unless not a single individual existed on the mother earth speaking that language.

Once the complexity of compositional language was well-assimilated by the human probably the original significance of finding the correspondences between the object or event and the sounds used for them was lost or blended beyond recognition but it sowed the initial seeds of lexicon, helping to form original vocabulary foundation on which subsequent linguistic elaboration was built.

In todays language there are many words, related to the new introduction of gadgets and the words intodays language are related to those gadgets such as computer related words like download, hardware, soft ware, etc., or cultural influence such as introduction of many rituals or games, with cricket came the words like, keeper, bowler, etc., or the new words coming in use after mobile culture such as 'u' for you and 'r' for are, may we call it a mobilanguage?

What Happened to Hmmmm?

Homo sapiens had words followed by syntax and semantics. Then what happened to Hmmmm vocalization of pre-human species? Hmmmm had manipulative, multimodal and mimetic functions. As all these functions were more efficiently and better served by newly acquired compositional language, as per rule of natural selection, all these functions, which were defunct now, disappeared when better alternative was available. The production of phonetic symbols went to left brain instead of right as in Hmmmm calls, phonetic symbols—the phonemes were produced by neural circuits for movements of articulatory organs which were present in left frontal cortex. there were already neural circuits present for the functions related to chewing, sucking, and eating, present for millions of years. There was exaptation (borrowing) of the same circuits for the new function of speech complete with gene duplication and exaptation. Therefore the functions which were taken over by newly acquired phonetic language also went to left brain. The activity, however related to the reproductive function in a way, to attract the female for mating was retained. This activity was music the fourth m in Hmmmm signal. One may think, was music related with communicative language? Yes, very much.

Without thinking of music we will be missing all the prosodic features and infant directed speech or IDS, of compositional language. Only difference was it was shifted from limbic system of reptile brain, which is a seat for production of vocal calls in pre-human species and mammalians, to right brain cortex. Right brain is specialized for all spatial functions such as fine art, fine skills, painting and music. Therefore music was shifted to right brain.

What happened to holistic character of Hmmmm? Alison Wray has published her book "Formulaic Language and the Lexicon" in 2002. She advocates, that even though we have now the compositional language, holistic language is an important part of it. We often use proverbs, or what can be described as formulaic phrases which take help of metaphor. Metaphor is an attribute exclusively available to human only. The phrases in common communication language such as "come on, let us go to the table", means meals are ready, served and we will go to the table to eat. Proverbs are more used by the people living near the nature, such as living in rural atmosphere. We often see the famous Marathi actor Makarand Anaspure using proverbs in his dialogues. When he says "he is like a jackfruit" it metaphorically means, though 'his speech is apparently thorny, but at heart he is very sweet and affectionate like fruit of jackfruit tree.' The terms like "cool" "awesome" popular in modern youths speak a lot than the words apparently mean.

We can say that these are the remnants of holistic calls of pre-human species, which remind us of Eernsts Haeckel's theory of onto genic recapitulation or phylogeny.

Neurological Evidence

After our discussion about, origin of language and claim that, it is based on logical thought, a question will be raised is there any neurological evidence to prove your hypothesis? Yes, it comes from various clinical cases seen by neurophysicians.

If the disease site is IPL, supramarginal and angular gyri may be affected. If the supramarginal gyrus is affected, the patient suffers from apraxia. His mental condition is good including his comprehension and production of language. But if you ask him to imitate an action he fails. He understands what to do but is unable to execute. He cannot even judge if someone else is doing it correctly or not. Their problem lies in linking perception and motor ability. They are completely unaware that they are miming the act incorrectly. In case of playing a musical instrument, say piano he knows how to play but fine movements of fingers are lost and he cannot play.

If there is failure in the angular gyrus of IPL there is anomia. They have difficulty in naming even the common objects. They know everything about the object, even its function, but fail to name it. They may substitute a related word such as cow for a pig. They fail in interpreting proverbs. They will tell the literal meaning but miss the metaphor. If you say 'a rolling stone gathers no moss' 'they will tell you' if you see a big stone coming down a hill it will not collect any dirt with it.

If there is fault in Wernicke's area, the semantic matter will reach the Broca's area and will be organized properly in syntactic arrangement, as there is no fault in Broca's area but the utterance will be a gibberish that is without any meaning.

If there is any fault in Broca's area the semantics coming from the Wernicke's area will not be organized as per syntactic rules, and will be completely disorganized or there will be no motor action at all.

These are by no means the only parts of the brain where the damage can lead to particular clusters of disruptions: For instance individuals with damage to the arcuate fasciculus, which links Broca's and Wernicke's area, often suffer from a condition known as conduction aphasia (though some conduction aphasics have damage elsewhere instead). Conduction aphasia involves, difficulties in repeating words, along with a high frequency of phonological errors, notably substitution of sounds within words. It has been suggested (Obler and Gjerlow 1999) that this implicates arcuate fasciculus is also involved in phonological planning, where phonemes are put together in the right sequence to make words.

Genetics of Language

The detail study of genetic foundation of 'the origin of the language' or the language use in it is not yet complete. We know that all the physiological functions executed in the body are through specific neural circuits, these neural circuits have got genetic backing. In short almost all the activities in the body are functioning under the influence of genes (innate).

Is the language, also under the genetic control? Yes, if specific genes are defective the language is also defective.

Myrna Gopnik working in McGill University has studied a family named KE. She had seen a television show in UK. Produced by Vergha-Khadem and her colleagues. Gopnik studied four generations of this KE family in detail. She found that 50% of the members of this family were having defective language. They had difficulty in comprehending past tense, which was syntactically correct. She would not understand regular past tense such as walk-walked but had no difficulty with irregular ones like 'go-went.' One of the reason for finding defective past tense is because the regular past tense forms are processed by a neural circuit that includes the left superior temporal gyrus, Wernicke's area and connections to the left inferior frontal gyrus. Irregular verbs however take a different path through the brain. It appears as if the stem and the affix of the regular past tense verbs are computed as the words with difficulty to be heard, but the irregulars, which have no special syntactic marking, are treated simply as whole words, like nouns or uninflected verbs. Vergha-Khadem had found many more defects in their language. Their articulation was also defective and render the speech unintelligible, she had to take help of subtitles for her television show. They were unable to understand singular and plural. If you showed them one apple they could not count two apples or many apples. Their facial gestures were also faulty. The IQ of the family members was also on an average, less by 18 to 19 points.

In 2001, genetist's found out that, their FOXP2 gene was defective. Gene is formed by nucleotides made up of amino acids, this is not a gene very specific for human. It is also present in chimpanzee and rats, where, out of 700 pairs of amino acids two pairs are different in chimpanzee and 3 pairs in rats. In rats the gene might be performing different function as rats have got no language.

It was shown that the fault in the KE family was due to defective FOXP2. This defect is known as specific language impairment or SLI.

It is not certain that FOXP2 gene is related in any way to the origin of language. It might be functioning with other genes hand in hand or it might be stimulating other genes. There is much to be still known about genetic influence on language. FOXP2 might have been changed as a result of mutation in tandem with the evolution of *Homo sapiens*.

Semantics and Syntax

I have discussed up till now everything about a word, how they are formed, their characteristics, utterance, etc. Utterance of phonemes and words is a unique step, but language is not just words. There are two other important aspects to be considered, syntax and semantics. How are these represented in brain and how did they evolve? These two functions are autonomous. Without syntax and semantics we would have got only words to communicate with each other, again a holistic type of language. But it is not so. The words are grouped in meaningful matter or semantics in Wernick's area. This is taken over to Broca's area where the matter is organized in sentences as per syntactic rules. These two functions, namely syntax and semantics are therefore not dependent on each other. If Wernicke's area is faulty Broca's area will still juggle the faulty matter reaching it from the Wernicke. The words with correct grammatical rules in a meaningless gibberish manners. There will not be any complex or recursive sentences, as this is the function of Wernicke's area.

Let us look at **semantics.** What is semantic? A meaningful sentence. What is meaning? We know that the function of the Wernicke's area is comprehension. Comprehension is what we understand of, what reaches to auditory center in temporal lobe in left brain? We listen the speech and language in the form of sound waves reaching the ear. These sound waves are transformed into congruent electric waves in the inner ear. These electric waves travel via auditory nerve in the brain and ultimately reach the auditory center in the temporal lobe left side. Wernicke's area is adjacent to auditory center. In Wernicke's area the electric wave is decoded, what we call as comprehension. Decoding of language heard is dependent on our previous knowledge of association of words and what they stand for? The word is of course an arbitrary symbol to a physical attribute of sounds related to an actual object or action. The relation is accepted by the society and becomes a part of the language. This relation between the sound and the object or the event is nonarbitrary as we have seen. When you hear the same sound, which forms a pair with that object or event, Wernicke's area understands the meaning of the sound, that is what it stands for and the word affixed to it. This is comprehension or decoding of sound wave. That is perhaps what is meaning. In fact we have no idea how neurons in this area actually do their job. Indeed the manner in which the neural circuitry embodies meaning, is one of the great unsolved mysteries of the neuroscience. But in our understanding the nature of comprehension of auditory sensation, the initial abstraction between sound-shape/size translation is accepted as basis.

Wernicke's area is a part of IPL. IPL is one of the most important junction or the crossroad as described previously between touch and joint sense, vision, and hearing. It is easier here to have abstraction of one sensation to other or synesthesia, to take place in the presence of mirror neurons. Here visual profile of an object is abstracted to an auditory wave it stands for to which we have given an arbitrary symbol. Abstraction is a function unique to human.

IPL is a part which starts growing in size in primates as they need more coordination in two senses vision and joint sense. They have to negotiate branches of a tree for climbing up and that to achieve they require accurate knowledge of branch, its size, direction, etc. and joint sense. That is why visual sense is abstracted to joint sense.

In human there is accelerated growth in IPL. The lower part splits in humans, probably as a result of gene duplication, a frequent occurrence in evolution. The upper part, the supra-marginal gyrus retains the old phylogenic function of hand–eye coordination, elaborating it to the new level of sophistication required for skillful use of tool manufacturing, skilled use of fingers and joints for carving and art, threading a needle and also imitation, in humans. In angular gyrus the very same ability is abstracted for naming the objects from visual profile. This is what is decoding sound in naming?

We have already seen that IPL area is rich in mirror neurons and is ideally suited for the job of abstraction of one sensation to other. The accelerated growth of IPL and its splitting in two must have played a pivotal role in the emergence of functions unique to human. Those functions include high level abstractions. Since we have two angular gyro, one on each side of the brain, they are specialized in two different types of abstractions. One on left for temporal events like language and metaphors related to language and on the right for visuospatial abstractions including music, arts, etc. and body-based metaphor.

To prepare a reply for someone else's talk to us, or parting information from memory is the function of Wernicke's area. One more function of the Wernicke's area is abstraction from one module of intelligence into other and compose a thought-based on inter-modular intelligence. As described previously in evolution human has got three types of cognitive modules, or intelligences as it is called by Peter Caruthers. The natural history intelligence, or knowing everything related to hunting and surviving. Second is social intelligence for keeping social relations, informing each other about danger of predators, and other natural calamities and about life after death, idol worshiping, etc. And the third technical intelligence for manufacturing tools, various arts and crafts.

To come to the language once more, human has got capacity to integrate, one intelligence into, other type of intelligences and create embedded or recursive (complex) sentences and thought. The syntax part of the compositional language is, one sentence generated by one type of cognitive module/ intelligence to be embedded into that of another sentence coming from a different module/intelligence. Thus, a single imaginary sentence would be created which would be an inter-modular or cognitively fluid thought. This could not have existed without compositional language. Thus, creating recursive sentences is a part of semantics and is the function of Wernicke's area. See for example the following sentence "I saw the young girl, your girlfriend, in City center mall, selling dresses on the counter in a shop, she was beautiful, having the same name as told by you, therefore I gave the book to her".

Thus, subject matter assembled in Wernicke's area as a reply to someone else's statement or imparting information about one's thoughts, taking words from memory, and forming recursive sentences whenever necessary, is all what is semantic. Semantic is transferred to Broca's area via arcuate fibers for further processing to syntactical structures for uttering.

Syntax

In the section of syntactics, I have taken help of Dr Ramachandran's thoughts also. Let us now turn to syntax or syntactical structure. At the basic, syntax is a series of rules for combining words in a meaningful sentence. This gives to human language enormous range and flexibility. Are the syntactic rules innate or exapted from rules evolved for some other function? Semantic language before it is expressed as speech via articulatory system, is organized in the form of sentences or syntactical structure. This includes co-localization of circuits required for motor planning and language processing in the form of syntactic arrangement or a sentence. Here are already motor action circuits for different actions present from the pre-human species. These circuits form a basis and are exapted for other function, namely formation of, or organizing words in the form of syntax formation for utterance through articulatory system. That is a consequence of exaptation of language structures from a preexisting brain system developed to plan and control complex motor sequences and in particular some actions like throwing (which is easiest to acquire) or breaking a stone or a hard fruit or cutting.

Arbib says only simple actions can be copied which have persisted with our common ancestors to pre-human species to human. They can provide a scaffolding for nonverbal communication-proto-sign, which provided a pathway for proto-speech and syntax. These are ideas of adapting and returning preexisting components of the brain (and indeed other relevant systems).

Formation of syntax is the function of Broca's area. Broca's area is a part of motor cortex in front of the central gyrus—area 44/45 of Brodmann. It is just in front of motor area represented by jaw and mouth organs. Broca's area is newly evolved part of frontal cortex found only in human. Compositional language is to be abstracted from auditory sensation into motor maps in Broca's area before it is spoken via articulatory system.

The motor area in pre-central gyrus is already functioning for many other functions in pre-human species and mammals like in apes since a few millions of years, such as expression by gestures. Also for motor functions such as activities related to survival instinct. Think of an action like cracking a hard fruit like coconut. The sequence of events is already there. One who is cracking or in action (subject) the action he is performing, cracking (verb) and the object on which the action is performed (object). Splitting of neural circuit from the original neural circuit into two, which might be associated with gene splitting is an event often occurring in evolution. Out of the two circuits thus evolved the original retains its function, while a new function is exapted from the old one for the new circuit. The new function is related to language for organizing a syntactical structure such as subject-verb-object. This thought can be elaborated in respect of, more complex functions already existing. The original function is entirely dissimilar to the new one. More than one things happening such as instead of a fruit, cracking a stone, for preparing a tool with a sharp edge and fixing it on a handle for preparing an axe. There are two different functions combined into one for having better weapon. This function is exapted by the new circuit for having a compound sentence. Or fixing the stone chip on a pole, and extending the length, by fixing one more pole to it. This function is exapted for combining more than two sentences in a compound sentence having more than one clauses. Does the original function is affected when the neural circuit is split into two? No, the original function continues as it is, as long as it is beneficial to the species. If the function of Broca's area is affected by disease, the semantic matter received from the Wernicke's area will not be organized in syntactical structure even if it is forwarded for articulation and it will be unintelligible. Most of the time if the motor area is affected by the disease there will not be any articulation, resulting in aphasia.

This is what is syntax formation a function of Broca's area?

Evolution does not take place with any intention. Variations appear in certain individuals of a species and if they are found beneficial, for the better survival and reproductive process. they are carried further in next generation. In few generation they become genetic and a new structure and/or function is evolved.

Human speech involves a number of anatomical and neural mechanisms. The plurality of these mechanisms is consistent with mosaic nature of evolution which takes place by bits and pieces and therefore might be genetically transmitted. As they are added by bits and pieces we cannot say that they are specifically evolved for a particular function say language. At the end it might be a structure which fits in the mosaic picture for a particular purpose. Richard Dawkins in his book "The Blind Watchmaker Quotes one most famous passage from 19th century Theologize Paley." In crossing a road suppose I pitched my foot against a stone, and were asked how the stone happened to be there? I might possibly answer, that for anything I knew to the contrary it had lain there forever. Nor would it be possible to show the absurdity of this answer. But suppose I had found a watch on the ground, and it should have been there? I should hardly think of the answer I had given before. For I knew that the watch must be there but in case of watch it must have had a maker, there he must have existed at sometime some artificers who gathered pieces from places and formed it for the purpose who comprehended its construction and designed its use and for what purpose was decided later (edited by author)?

In the evolution of speech some such thing must have happened. Structures were there, their functions were adapted or abstracted, or exapted (borrowed) for entirely new function. The very broad example of this we know is of, organs of articulation which were originally evolved for swallowing, cutting, chewing food, etc. were exapted for the function of speech. Let us take a review of evolution of speech and language from this point of view.

In the IPL visual muscle and joint senses, were the two sensations used for the fine control and fine actions of fingers for holding the branches of the tree in apes. In the same region, in the presence of mirror neurons by adaptation or abstraction parallel was found between the two the visual and auditory sensations. The structures were there only the function of adaptation between the two different sensations, in place of joints and visual now between visual and auditory took place. The auditory wave was already generated in pre-human species for their vocal calls now it is adapted for visual and auditory sensations for more precision. The wave is carried as before to Broca's area for stimulating articulatory system for utterance.

The neural circuits for gestures were already there in area 44 (future Broca's area) in the apes evolved more than 20 million years ago for gestures. The apes were accompanying there communication with gestures for which they were using the area 44 of Brodmann. The area 44 is the precursor of future Broca's area in modern human. As we have seen before Robin Dunbar says communication in apes was with personal grooming. When the groups enlarged the personal grooming became difficult and the grooming was transferred to vocal grooming. Thus, adaptation of visual sensation for grooming and gestures was already taking place. The grooming was now by gestures and vocal messages. In case of human however the vocal communication was shifted through articulatory system to compositional language. The articulatory system was stimulated by auditory wave and therefore the neural circuits from visual to gestures for having gestures parallel to visual sensations were split into two one for retaining old function of production of gestures and new one for adaptation of auditory wave, parallel to the exact size of the object in the presence of mirror neurons. Structure was there only it was duplicated for new function. In the articulatory system the organs were there for the production of vocal calls only there was some change in the dimensions and positions of the structures such as jaw receded back. The receding of jaw was gradually taking place in pre-human species because of evolution of neocortex. The frontal lobe grew in front in prefrontal area and forehead was pushed forward and the skull was flexed only in modern human. Tongue and larynx which were already moving down because of bipedal position moved still more thus a proper supraventricular vocal tract or SVT was formed only in human of perfect shape not for linguistic purposes but because of achieving perfect biped position. The thin anterior two-thirds part of tongue remained in the mouth which was more agile, and by touching the palate, could produce variety of vocal utterances called phonemes. Phoneme is a physical attribute of sound with their linguistic functions. By combining phonemes in complex units by permutation and combination, human could increase his repertoire or vocabulary and that was the evolution of speech and latter language. This is in short the evolution of speech I have put omitting details. In this we cannote that no new structure or function evolved with the intention of developing speech but human has taken advantage of the existing structures and there functions, modified or exapted wherever necessary for his cultural evolution of compositional language. The evolution of speech and language is therefore basically cultural evolution based on adaptation or abstraction and exaptation of some functions and structures which were already there. This is what Charles Darwin wrote in his book "Descent of Man" that language is half art (culture) and half instinct (innately present structures and functions adapted).

SUMMARY

The sound wave travelling through the ear, cochlea, getting converted in an electrical wave reaches brainstem. From there it travels to the auditory center in left temporal lobe, then reaches Wernicke's area and is comprehended, and is then either answered or stored in memory. It is its last station. It is logical therefore when you want to speak, a new wave is generated in the Wernicke's area only. Wernicke and angular gyrus are parts of IPL and here the auditory wave is translated or adapted parallel to the visual profile of object or action received via visual sensation received by angular gyrus. This takes place in presence of mirror neurons. From here auditory wave reaches the Broca's area. Neural circuits are already present there for gestures since millions of years. They split and by exaptation of function render auditory wave exactly matching to the visual profile of the object or action, in the presence of mirror neurons. Added prosodic effects by neural connections from basal nuclei, amygdala and anterior cingulate gyrus which are part of reptile brain developed already in reptiles. A goal-oriented stimulus from Broca's area reaches the brainstem and a phoneme having formant frequency needed for its utterance is uttered by proper coordination of muscle group competent for the utterance of that phoneme. Sequentially uttered vowels or consonants form a word. Words make sentences and then semantics. It is a cultural origin or evolution making use of structures already evolved for some other purpose or are still performing those purposes. In this theory there was no need of taking help of any innate factor such as language organ, universal grammar or language acquisition device (LAD) or even an abstract mental space.

Furthermore a highly complex set of behavioral or cultural system like human language are likely to be influenced by many genes not by only one. Mutations will typically be observable only rarely when their effects are unusually large and disrupt many systems.

This is the complete story of origin of our faculty of speech and language. This is the special attribute that distinguishes modern human from all other species existing or existed before on the globe.

Flowchart 10.1: Word for large object or high speed

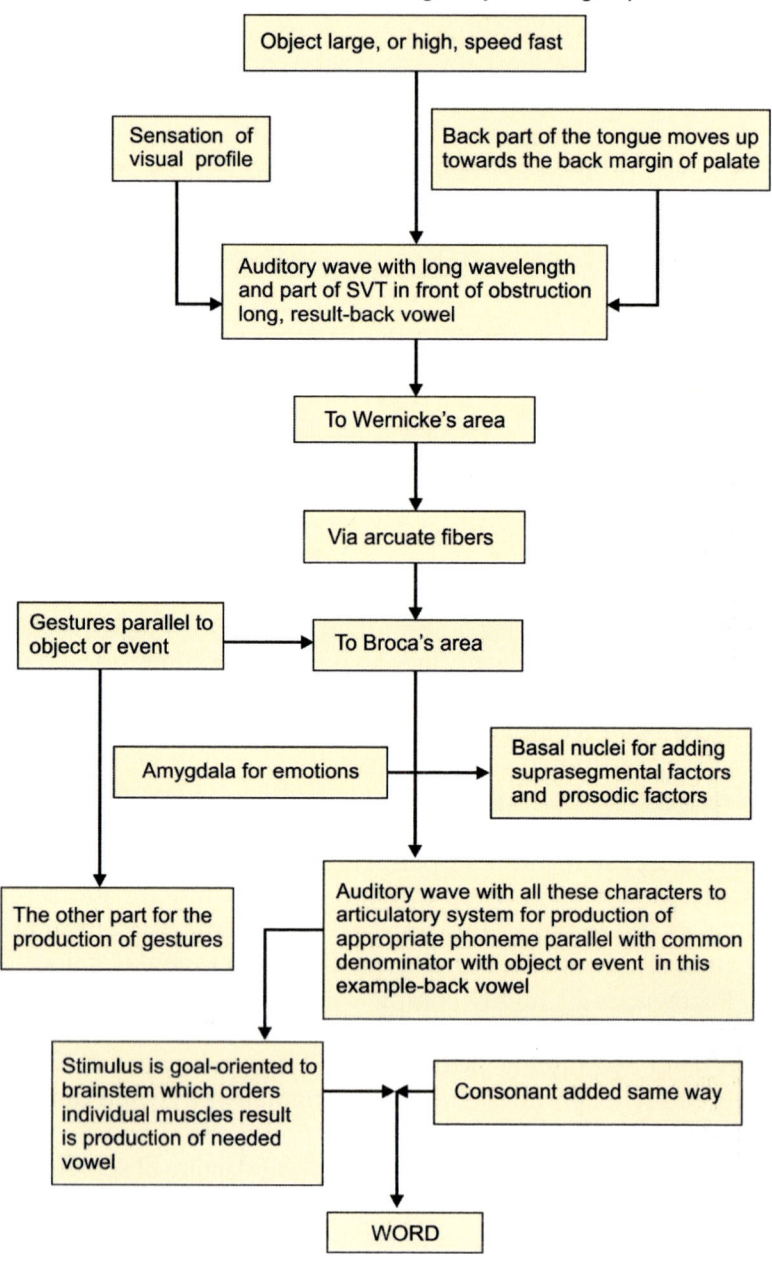

Flowchart 10.2: Word for small object or slow speed

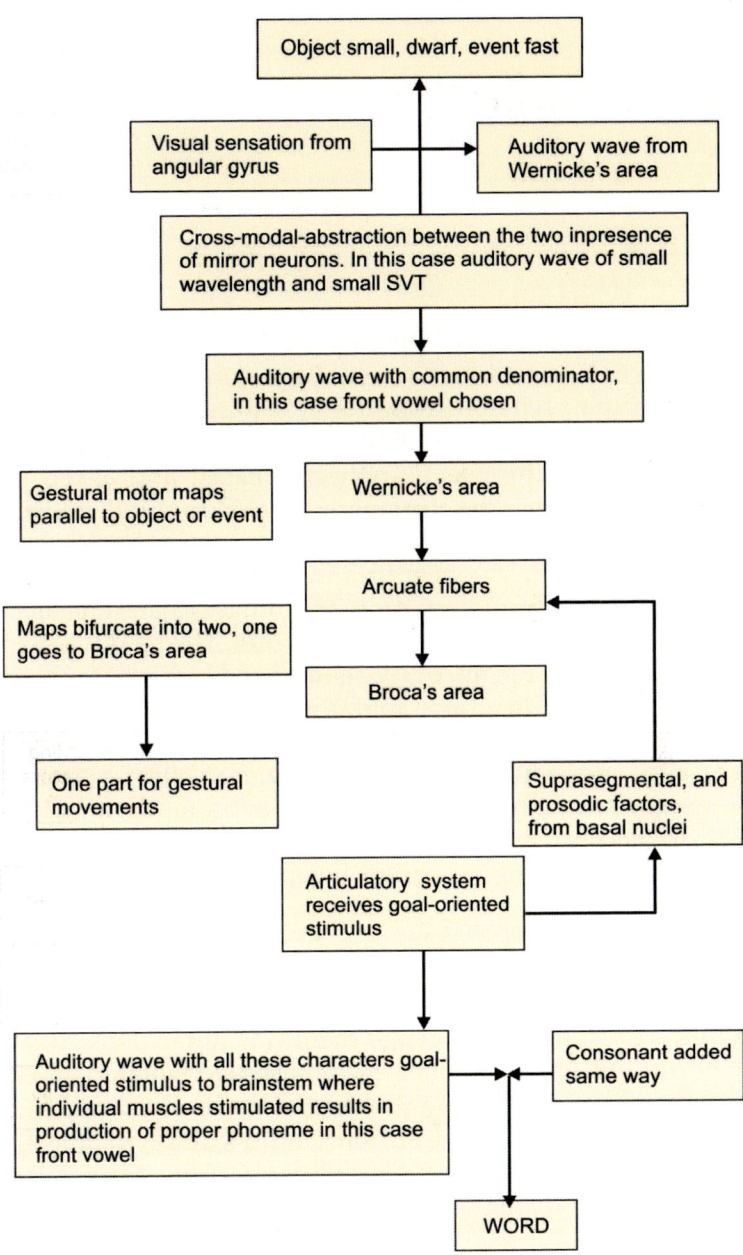

11 Epilogue

I just remember a story from famous 'Fables of Aesop'.
Once a Sultan told his favorite Khansama (chef) "Today I want a recipe prepared from the best things in the world". Khansama prepared a very juicy and tasty recipe for him . After tasting the food he was extremely pleased and asked the Khansama what was the chief ingredient? He humbly told him "Sir, I have prepared it from the tongues of the animals". Next day Sultan had a very queer idea in his mind. He told him to prepare a recipe from very ordinary things. Khansama again prepared a very tasty item and served him. The food was very tasty and he was delighted. He again asked him what was the ingredient? And Khansama told him "Sir, today's recipe was also prepared from the tongues of the animals." Sultan was very, very angry and scolded him "You are a cheat, yesterday the tongues were the best things, and how come, they became ordinary and bad in one day?" Khansama very softly told him "Sir, tongue is never good or bad. It depends on how we use it".

Why Language Evolved?

Did language evolved with some intention? In evolution nothing happens with any intention, we have already seen that. With the evolution of language there was miraculous change in the lifestyle of the human race. Man started wondering why this miracle of language might have been gifted to him?

We have already seen Stiffen Gould's thought that the language was not evolved for communication. Man started thinking first and language evolved as an accessory to express the thoughts.

Some think that language evolved for having mutual dialogue between each other for manufacturing better tools, better strategies for hunting, and giving calls about predator attacks, Charles Darwin in his book "Descent of man and selection of sex" writes that language and music evolved as an attraction to the mate for better chances of reproduction.

Whatever might be the reason but language is a special attribute of the *Homo sapiens*. Man has infinite number of vocabulary, and variety of subjects for mutual discussion. *Homo sapiens* migrated out of Africa between 80,000 and 100,000 thousand BP and thereafter. In a relatively short span of 50,000 years, there was very rich cultural, and spectacular progress, in scientific field. Instead of single domain cognitivity in pre-human species *Homo sapiens* had interdomain cognitivity (cognitive fluidity).

Art of Writing

Writing came long time after the evolution of language. It did not evolve by any intention of exploring something new, not that *Homo sapiens* one day thought let us now explore how to write this language that we speak? They migrated long distances group by group and then it became difficult to keep contact with each other. They had already developed many skills and arts and there was great progress in scientific field also. Art of growing food by agricultural skills was already developed by 10,000 BP. They had started sending artistic pieces and food grains from one place to another and it became a necessity to keep a record of goods thus exchanged. The flow of goods was from east to west specially to middle east countries in the beginning. It became necessary to keep a record of from where the goods came, from whom and how much? They did this by putting identification marks in the form of pictures of animals, such as bulls, sheep, cow, etc. and birds and some signs on clay tiles. This was the first stage of art of writing. This was actually a pictorial script and is called 'cuneiform.' This beginning of writing was somewhere between 6000 and 7000 BC.

Advantages and Disadvantages of Language Evolution

There is no need of talking anything much about advantages of this miraculous gift of language to mankind, we are all enjoying them. But can there be disadvantages of acquisition of such a gift of evolution. Yes, there are some. The great surgeon and anatomist of 20th century Sir Victor Negus has written a book on "comparative anatomy of larynx and pharynx" in 1949. In case of animals the vocal cords are comparatively thick and they move away more during inspiration, the shape of their larynx is suitable for that. That way they can inspire more air in, which they need as they often need to run away fast and long distances to save themselves from predators. This is the need of their life and survival. In *Homo sapiens* the cords have become thin, the reasons for which we have already seen. The cords do not move long distance, as they have to return back quickly for speech. Relatively man can inspire less air during the act of inspiration which compromises his physical fitness.

The primates and pre-human species have got prognosthesia or their jaw is forward projected. Naturally there is enough space for the third molar to erupt. In case of human the jaw is receded back and the space behind the second molar is very much compromised and the third molar cannot erupt straight up. It tries to get erupted in a crowded space. There is swelling and inflammation and there can be bacterial contamination. It has to be removed by surgical procedure. When anesthetic drugs were not invented this surgery was very difficult and painful. Sometimes the condition used to be fatal because of contamination. The man has started eating more and more soft cooked food and there is no need of so many molar teeth and the third molar germ may not be there for eruption. Man will have then only 28 teeth. The writer of this treatise is a living example as I have got only 30 teeth.

One more disadvantage is due to our bipedal position. It is our observation that the infant has not to stop in between while sucking for breathing as its larynx is situated high up just behind the tongue like primates, and while sucking it pulls it up behind the nose and locks so that it can breath and suck at the same time. When the man acquires complete bipedal position the vertebral column is pushed forward and the larynx shifts downwards in front of the food passage. It becomes necessary therefore to close the larynx tightly while eating. This happens by reflex action. This reflex action develops when the child is growing and starts sitting by the age of six months to one year. In the beginning when the reflex is not matured, accidentally larynx remaining partially opened, food particles may enter the larynx through the wind pipe and even reach lungs causing obstruction to breathing. This can happen even in adults some time while eating food in a hurry, or trying to eat and speak at the same time. Sometime in parties one or two shots more and this can happen in adults too, even deaths are recorded after such an accident.

Genetic Continuity?

Noam Chomsky stated that 'there evolved a language organ in *Homo sapiens* and the syntax rules and language are innate genetically transmitted in the human brain from the birth. Linguistic environment simply activates this universal grammar.' This was the belief of many contemporary linguist. Chomsky proposed a gene or genes to determine complex key aspects of human language. Similarly in his book 'The Moral Animal', Robert Write misinterpreted the evolutionary biology and proposed existence of a 'morality gene.' However, if the morality was actually regulated by a genetically specified organ then given the application of the principles of Mendel's laws there might be many individuals, their children and their children's children, entire family and community would inherently be immoral. This would present a moral dilemma. Morality clearly has a cognitive, linguistic and cultural basis and not genetic.

No one has ever identified genes for various human behavior like altruism or aggression, morality or immorality. The approach of adaptationist to human social behavior often called 'sociobiology' after a book by the same name by EO Wilson (1975). Our brains decision making mechanisms were shaped by natural selection to enhance fitness. Not to provide us with the capacity to monitor the consequences of each and every action. The individual is, and motivated to do things that lead to direct or indirect fitness gains. We enjoy sweet food, dislike tasteless, we fall in love, we derive satisfaction from charitable deeds, and we derive approval from others. We learn language because we possess physiological mechanism that make us want to learn it and enjoy benefits of communication with other individuals.

Moreover a behavioral ability is adaptive, it does not demand that the characteristic be 'determined' by genes. There are no alleles (minor variations of a gene) for altruism or aggression or any other trait, in the sense of a gene that encodes the characteristic and provides it no matter what behavioral development depends upon the interaction between the genetic recipe and the environment of the developing individual? Change the environment and you may well change the outcome.

A classic example in human biology involves language acquisition, a clearly adaptive ability for which there must be a genetic foundation. Among the many genes that we possess are some that code for enzymes that play role in the development of a brain with certain specialized components, like IPL and mirror neurons that facilitate language learning. But these genes for the language trait require all sorts of environmental inputs both in the form of neuronal development and acoustical experience. In order that the trait to appear in the individual, alter the environment and the diversity of language that gives a child different acoustic experience and the characteristics of the trait will change.

If morality was genetic Bhakta Pralhad a benevolent king in Hindu mythology would not have been born of demon Hiranyakashyapu. Another example is that of conqueror of the Asian land from east coast to west, capitulated the ancient city of Merv in 1221 and "Son of Jenghiz Khan Toloi seated on a golden chair in the plains of Merv, witnessed the mass execution. Men, women and children were separated distributed in herds and beheaded. But Mongol 'morals' clearly had nothing to do with Mogul genes. Kublai Khan the grandson of Jenghiz Khan who ascended the Mongol throne in 1260 was hailed as venerable divine by Buddhist sages. The Tibetan monks then converted the mongols to Buddhism. Marco Polo who visited the Kublai Khan court found that he founded monasteries, repaired the imperial roads, planted shade trees, and provided assistance for old scholars, orphans and the sick and infirm. Buddhist influence was so strong that rice and millet were distributed to poor families. Marco Polo journals note that Kublai Khan himself fed thirty thousand paupers daily.

Word and language is a medium for emotional expressions, thought and bridge to connect historical facts and values for future. At the same time it can create serious misunderstandings, confrontation amongst leaders and induce wars amongst nations too. Dr Manmohan Singh had a joint declaration with Pakistan in 2004. Just one word out of the manuscript created a great uproar in the parliament. Politicians critically do so much of juggling of words that the essence of the very debate was lost and there was only a battle of words amongst each other.

As Bernard Shaw says there is linguistic schizophrenia among different linguistic groups. TV channels make a capital of it and go on ruminating it to increase their TRP. Bernard Shaw further rightly says language enriched by values is the basis for civilization.

Depending on where you are born and in which house, depends whether you have to memorize the passages from Geeta, Koran, or Bible, and you will be indoctrinated by the philosophy of Shakespeare or Shankarcharya, Dnyaneshwari or Keshavsut. The medium to bring it to your conscious mind is language only. If language was not there we would have remained on the level of primate intelligence, only.

If the members of the society communicate in proper, meaningful, value based language, then only there can be public performances like dramatic activity, cinema, and literature of high cultural values. The words out of context and differences of religious faiths, may induce quarrels, and big confrontations may erupt which can cause big encounters that may lead to anhilation of races. The moral is, words are not good or bad, as taught by Aesop's story in the beginning of this chapter, proper use of it only can bring brotherhood and fraternity in the mankind.

I do not know whether language can be considered as a variable in the light of Darwin's theory of 'natural selection'. But as per principles of evolution the variables which are not beneficial to the species are lost. If we do not make use of this gift received by human as an exclusive attribute, properly, only to induce brotherhood and love amongst each other but use it only to induce a great confrontation amongst races and countries that may lead to mass destruction of civilizations. What will happen? Under such situations this language variable, may be proved unwanted and not beneficial to the species. In that case we may loose this attribute of language, and there will be no one to witness and record such an event, as only, 'we can talk'.

Bibliography

1. Aitcheson J. The seeds of speech; Language, origin and evolution, Cambridge University Press, Cambridge, England.
   'On toots of Language' Language and Communication; 3:83–97. 'The Articulate Mammal (1989).'
2. Atkinson JR. Aspects of intonation in speech. Implications from an experimental study of fundamental frequency, university of Connecticut, Storrs (1973).
3. Arbib Michael A."From Monkey Like Action Recognition to Human Language: An Evolutionary Freamework for Neurolinguistics"Behavioral and Brain Sciences" 2005;28 105–24. The Mirror System, Imitation and Evolution of Language in Imitation in Animals and Artifacts, Cambridge. 'The Mirror System Hypothesis, how did protolanguage evolved? In Tallerman (ed) 21–47.
4. Armstrong David F. (1999) Original Signs, Gesture Signs, and the Sources of Language, Washington DC. Gallaudet University Press , William C Stokoe and Sherman E Wilcox (1995). Gesture and The Nature of Language. Cambridge University Press, America.
5. Baldwin JM. 'A New Factor in Evolution. American Naturalist 1896;30:441–51.
6. Bates Elizabeth (2004). Explaining and interpreting deficits in language development across clinical groups: Where do we go from here? Brain and Language 1996;88:248–53. Learning Rediscovered. Science 1998;274:1849–50. "Construction Grammar and its Implications for Child Language Research." Journal of Child Language 25: 443–84.

7. Berger Doritha. Music therapy, sensory integration and the autistic child. London, Jesica Kingsley Publication (2002).

8. Berlin Brunt (1992). The Principles of Ethnobiological Classification. Princeton University Press 2005. Just another fish story, size symbolic properties of fish names, in animal names Venezia: Instituto Veneto D Sciences.

9. Bickerton D. Language and species. Chicago, Chicago University Press (1998).

10. Bickerton D. Catastrophic evolution from the step for a single step from protolanguage to full human language in approaches to the evolution of language and cognitive biases. Cambridge, Cambridge University Press (1998).

11. Bickerton D, (2000). How protolanguage became language in the evolutionary emergence of language? Cambridge, Cambridge University Press. Language and Species. University of Chicago Press (1990). Language Evolution: A Brief Guide for Linguistics. Lingua. 117:510–28. Language and Human Behavior (Seatle; University of Washington Press). Chomsky:"Between a Stony Brook and a Hard place" (2005).

12. Blacking J. How musical is man? Seattle: University of Washington Press.

13. Bloom L. Language development: Form and function in emerging grammars. Cambridge: MA: MIT Press (1970).

14. Bloomfield L. An introduction to the Study of Language. London Bell (1914).

15. Berlin B. The principal of ethnobiological classification. Princeton University Press (1992).

16. Berlin B. Just another fish story. Size symbolic properties of fish names in Animal names (eds A. Minelli. G. Ortaili and G. Singa) Venezia: Instituto Veneto D Sciences (2005).

17. Bishop Dorothy (2009). Genetics and environmental risks for specific speech and language impairment in children. Philosophical transactions of royal society of London. B:Biological sciences. 356:369–80. Language development in experimental circumstances.' London: Longman and K. Mogford (eds) (1988).

18. Broca Paul. Nouvelle observation d'aphemie produite par une lesion de la Motie posterieure des deuxieme et troisieme

circonvolution frontales Bulletine de la Societe d'Anatomique. Paris (1861).

19. Bond ZS. Identification of vowels excerpted from natural nasal context. Journal of the acoustical Society of America. 1976;59:1229–32.

20. Bruce D. Lashley and the problems of serial order. American psychologist 1994; 49: 93–105.

21. Brunet M, F Guy, et. al. A new hominid from the upper miocene of chad: Central Africa. Nature 2002;418:145–51.

22. Cantalupo Claudio and William D Hopkins. "Asymmetric Broca's area in great apes." Nature 2001;414:505.

23. Cavallis-Sforza. History and Geography of Human Genes. Princeton NJ (1994).

24. Carruther P. The cognitive functions of language. Brain and Behavioral Sciences 2002;25:657–726.

25. Chomsky Noam (1964). Current Issues in Linguistic Theory. The Hague Mouton (1966) Cartesian Linguistics, New York: Harper and Row (1986) knowledge of language its nature, origin and use. New York Praeger (1988): Language and Problems of Knowledge. The Managua Lectures, Cambridge (2000). The Architecture of Language, Oxford; Oxford University Press (1972). Language and Mind. New York: Harcourt, Brace, Jovanovich (1965), Aspects of the Theory of Syntaxs. Cambridge: MA:MIT Press.

26. Corbolis MC (1991). The Lop-sided Ape Evolution of the Generative Mind Oxford; Oxford University Press (2003) 'From Hand to Mouth: The gestural origin of language in Christiansen and Kirby.

27. Crelin ES (1969). Anatomy of the Newborn: An Atlas Philadelphia: Lea and Fibiger (1973). Functional Anatomy of the Newborn. New Haven Yale University Press.

28. Crick Francis. "What Mad Pursuit": A Personal View of Scientific Discovery. New York, Basle books (1988).

29. Cross I. Is music the most important thing we ever did? Music development and evolution in Music, Mind and Science Seoul, Seoul National University Press (2001).

30. Crow TJ (2000). Schizophrenia as the price that *Homo sapiens* pays for language: a resolution of the central paradox in the

origin of species. Brain research reviews 31:118–29 (2002a) Protocadherin XY: a candidate gene for cerebral asymmetry and language. In Wray (ed) 1–20:(2002b). The Speciation of Modern *Homo sapiens*. Oxford University Press (2002c). Introduction in Crow (ed)1–20 (2005) Who forgot paul Broca? The origins of language as test case for specification theory.' Journal of Linguistics 41:133–56.

31. Davidson I. The archeological evidence of language origin: states of art in "Language Evolution"Oxford: Oxford University.

32. Daniel Jones Outline of English Phonetics, New York, Phoneme, its Nature and Use. Cambridge, W. Haffer sons 1949.

33. Dissanayake E. Antecedents of temporal arts in early mother and infant interaction. In the origins of music, Cambridge, MA: MIT (2000).

34. Darwin Charles: The Origin of Species; WR Goya: The descent of Man Chicago. Encyclopedia Britanica: The Expression of Emotion in Man and Animal. London, John Murray.

35. Dawkins Richard. 'The Selfish Gene' Oxford University Press (1976).

    'The Extended Phenotype' Oxford University Press (1985)

    'The Blind Watchmaker' London Penguin (1986)

    'River Out of Eden' London, Harper Collins (1995)

    The Ancestors Tale. London Weidenfeld (2004)

36. Deacon TW. The Symbolic Species: The coevolution of language and Brain. London: Penguin (1972).

37. Diamond AS. History and Origin of Language, London, Methuen (1959).

38. Donald M. Origins of the modern mind. Cambridge, Harvard University Press (1991).

39. Dunbar RIM, 1996 Gossip Grooming and the Evolution of Language. London Faber and Faber to the 1998. Theory of Mind and the Evolution of Language in Approaches. Evolution of Language: Cambridge, Cambridge University Press 2003. The Origin and Subsequent Evolution of Language in Language Evolution. Oxford; Oxford University Press.

40. Dawkins R. The Selfish Gene Oxford: Oxford University Press (1976).

41. Descartes R. Discours de la Metodee (1637).

42. Eimas PO (1971). Speech perception in infants. Science 171.

43. Falk Dean. Prelinguistic evolution in early hominids, behavioral and brain science.

    Hominid brain evolution and the origin of music. Cambridge, MIT. Comparitive Anatomy of Larynx in Man and the Chimpanzee: Implications for Language in Neanderthal: American Journal of Physical Anthropology 1975;43:123–32.

44. Ferrein CJ. Memoires dacademie des sciences of des Paris.

45. Folkins JW and Zimmerman GN. Jaw—muscles activity during speech with the mandible fixed. Journal of the Acoustical Society of America 6(2000). Gazazaniga Michael S (2000). The New Cognitive Neurosciences, Cambridge: MA:MIT Press, Gazaniga NS. (1995). "The Cognitive Neurosciences." Cambridge, MA:MIT Press (2000). "The New Cognitive Neurosciences"(2nd Edition) Cambridge. MA: MIT Press.

46. Gopnik Myrha. Familial aggregation of a developmental Language disorder cognition.

47. Gould SJ and Verba ES. Exaptation a Missing Term in the Science of Form Paleobiology 8;4–15.

48. Gould Stephen J (1981). The Mismeasure of Man, New York, WW Norton and Co and Richard C Lewontin (1979). 'The Spandrels of San Marco and the Panglossian paradigm a critique of the adaptationist.

    'Programme' Proceedings of Hausser Marc D. The evolution of communication, the Royal Society of London. B 1996;205: 581–91.

    And ES Verba. Exaptation—a missing term in the science of Form: Paleobiology Cambridge: MA: MIT Press 1982;8:4–15.

    And CA Fowler. Fundamental frequency declination is not unique to human speech, evidence from non-human primates. Journal of the acoustical Society of America 1992;91:363–9.

    And W Tecumseh Fitch. What are the uniquely human components of the language faculty in Christiansen and

Kirby Noam Chomsky and W Tecumseh? Fitch (2002). "The faculty of language: What is it? Who has it? How did it evolved (2003)?

49. Hawkins John A. Explaining Language Universals. Oxford: Blackwell (1988).

50. Henshilwood CJ and Sealy J. Bone artifacts from the middle stone age at Blombos cave, Southern Cape, South Africa Current Anthropology 1997;38.

51. Humphrey N. The Color Currency of Nature Oxford: Oxford Huxley JM. "Collected essays of Thomas Huxley. Man's place in nature and other anthropological essays" Whitefish MT Kissinger Publications (1984).

52. Huxley TH. Collected essays of Thomas Huxley. 'Man's place in nature and other anthropological essays. MT Kissinger Publicity, 1871 on the relations of man to the lower animals, collected essays (1863–2005).

53. Ingman Max, Kaessmann, et al. Mitochondrial genome variation and the origin of modern human. Nature 408. The Archeology of human Origin (2000).

54. Issac G. The Archeology of Human Origin. Cambridge: Cambridge University Press.

55. Janker R. Fort Shritte Zur Rontgen Kinematographic Strahies 1931;44–658.

56. Jesperson Otto. Language of its Development and Origin London, Allen and Unwin (1922).

57. Jacobson R. Child Language and Phonological Universals. The Hague Mouton (1968).

58. Jacobson R and Waugh LR. The Sound Shape of Language Bloomington. Indiana University Press, 1979.

59. Jackenndoff Ray. Foundations of Language, Brain, Meaning Grammar Evolution (New York). Oxford University Press, 2002.

And Fred Lerdahl. "The Capacity for Music: What is it? and what is special about it"? Cognition, 2006.

And Steven Pinker. "The Nature of the Language Faculty and its Implications for Evolution of Language. Cognition.

Possible stages in the evolution of language capacity. Trends in Cognitive Sciences 1999;3:272–9.

60. Johanson Keith. Acoustics and Auditory Phonetics. 2nd edition. Oxford: Blackwell (2002).

61. Kirby Simon and Morten H. Christiansen 'From language learning to language evolution. In Christiansen and Kirby 2003;272–94.

    M Dowman and TL Griffiths. Innateness and culture in the evolution of language. USA 2007;105:10681–6.

62. Kuhl PK. Speech perception in early infancy. Perceptual constancy for specially dissimilar vowel categories. Journal of the acoustical Society of America 1979; 66: 1668–79.

    Early language acquisition: Cracking the speech code. Nature Reviews: Neuroscie 2004;5:831–43nc.

63. Kratzenstein CG. Sur la naissance de la formation des voyelles. Journal of Physiology 1780;21:358–81.

64. Kay RF, et al. The hypoglossal canal and the origin of human vocal behavior. Proceedings of the National Academy of Sciences 95a.

65. Ledgeford Peter. Elements of Acoustic Phonetics. 2nd Edition, University of Chicago (1996).

66. Levelt W JM. Speaking from Intention to Articulation. Cambridge; MA: MIT Press (1989).

67. Lieberman P DE and RC McCarthy. The ontogeny of cranial base angulation in humans and chimpanzee and its implications for reconstructing pharyngeal dimensions. Journal of Human Evolution 1999;36:487–517.

    On the Origins of Language: An Introduction to the Evolution of Human Speech New York, Mac Millan (1975).

    The Biology and Evolution of Language, Cambridge: MA Harvard University Press (1984).

    Current views on Neanderthal speech capabilities. A reply to Boe, et al. Journal of Phonetics (2007).

    And ES Crelin. On the speech of Neanderthal man. Linguistic Inquiry 1971;2:203–22.

    Vocal tract limitations on vowel repertoires of Rhesus monkeys and other non-humean primates. Science 1969; 164:1185–7.

Human Language and Our Reptilian Brain. The Subcortical Bases of Speech Syntax and thought Perspectives in Cognitive Neurosciences (Cambridge, Mass:Harvard University Press) (2000).

"On the Nature and Evolution of Neural Bases of Human language." Year Book of Physical Anthropology 2002;45:36–62.

"The evolution of Human Speech: Its Anatomical and Neural Bases." Curreny Anthropology 2007;48:39–66.

68. Linineaus Carbolus. Systema Naturae Holmiae Laurentius Salvius Vailable on line a (1758).

69. Lund JP and Kolta A. Brainstem circuits that control mastication: Do they have anything to say during speech Journal of Communication Disorders 2006;39:381–390.

70. Luria AR Aphasia in a Composer Journal of neurological Science 2;288–92.

71. Macdermot, et al. Identification of FOXP2 truncation as a novel cause of developmental speech and language deficits. American Journal of Human Genetics, 760; 74–80.

72. Maess B Etal. Musical syntax is processed in Broca's area, an MEG study Nature Neuroscience 2001;4:540–5.

73. McCrone J. The ape that spoke: Language Origin and Evolution, William Morrow, New York, p. 91–1991.

74. MacNeil D. Hand and Mind. What gestures Reveal About Thought? Chicago, Chicago University Press, 1992.

Language and gestures, Cambridge: Cambridge University Press (2000).

75. Maggi-Cecchi J and Collard M. A fossil stapes from Sterkfontein, South Africa and the hearing capabilities of early hominids. Journal of Human Evolution 2002;42:259–65.

76. Mania D and Mania U. Deliberate Engraving on Bone Artifacts of *Homo erectus*. Rock art Research 1988;5:91–107.

77. Metz-Lutz MN and Dahl E. Analysis of Word Comprehension in a Case of Pure Word Deafness. Brain Language 1984;23:13–25.

78. Miller LK. Musical Savants: Exceptional skill in the mentally retarded. Hills dale NJ Lawrence Erlbaum (1989).

79. Muller F Max. Lectures on Science of Language. London, Longman (1891).

80. Muller J. The Physiology of the Senses, Voice and the Muscular Motion with the Mental Faculties (w. baly. trans) London, Walton and Maberlay (1848).

81. Maclarnom Hewitt. The Evolution of Human Speech. The Role of Enhanced Breathing Control. American Journal of Physical Anthropology.

82. Marcus Garry F and Simon E. Fisher "FOXP2 in Focus: What can Genes Tell us About Speech and Language"? Trends in Cognitive Sciences 2003;7:257–62.

83. Negus Victor E. The comparative of Anatomy and Physiology of Larynx. New York, Hafner (1949).

84. Nettle Bruno. The Study of Ethnomusiology: Twenty-nine issues and concepts, Urban, IL, University of Illinois Press, 1983.

85. Nettle Daniel. Linguistic Diversity, Oxford University Press (1999).

86. Paley William (1802). Natural theology or Evidences of the Existence and Attributes of the Deity Collected from the Appearances of the Nature: 2nd edn. New York, Harper and Brothers (1847).

87. Paget R. Human speech. London Kegan Paul (1930).

88. Parsons L. Exploring the functional neuroanatomy of music performance. A Cognitive Neuroscience of Music Oxford, Oxford University Press (2003).

89. Penfield W and Jesper H. Epilepsy and the Functional Anatomy of the Human Brain. New York: Little Brown (1954).
    And Robert L. Speech and Brain Mechanism. Princeton NJ, Princeton University Press (1959).

90. Peretz I. Brain specialization for music: New Evidence From Congenital Amusia in Cognitive Neuroscience of music Oxford: Oxford University Press (2003).
    Etal: Congenital Amusia, a Disorder of Fine Grained Pitch Discrimination. Neuron 2002;33:185–91.

91. Peterson G. The phonetic value of vowels. Language 27 (1951).

And Barney HL. Control methods used in the study of vowels. The Journal of the Acoustical Society of America 1952;24:175–84.

92. Pilbeam David. New hominid skull material from the Miocene Pakistan (1982).

93. Pinker S and P Bloom. Natural language and natural selection. Bahavioral and Brain Sciences 13(4);707–84.

And Ray Jackendoff. The faculty of language. What special about it? Cognition 2005;95:201–36.

How the mind works (New York, WW. Norton 1999).

Talk of Genetics and Vice-versa. Nature 2001;413:465–66.

Words and Rules, an Exchange" The New York Review of Books 49 (2002).

Posner M. Foundations of Cognitive Science. Cambridge; MA: MIT Press (1989).

94. Rizolatti Giacomo and Destro MF. Mirror Neurons Scholar Pedia 2008;3(1)2055.

95. Rizolatti Giacomo and Michael A. Arbib. 'Language within our grasp.' Trends in Neurosciences 1998;21:188–94.

96. Saffran Jenny. Statistical learning by 8-month-old infants Science 1996;274:1926–28.

97. Saffran Jenny and GJ Griepentrog. Absolute pitch in infant auditory learning: evidence for developmental reorganization. Developmental Psychology 2001;37(1):74–85.

98. Sloboda JA. and Howe. Biographical precursors of musical excellence: an interview study. Psychology of Music 1991;19:3–21.

99. Saussure F De. Course in General Linguistics Philosophical Library (1959).

100. Schwartz DA CQ. Howe, and D. Purves. "The Statistical Structure of Human Speech Sounds Predict Musical Universals." Journal of Neuroscience 2003;23:7160–68.

101. Sidenberg MS. Language Acquisition and Use: Learning and applying Probabilistic Constraints Science 1997;275:1599–1603.

102. Sloboda JA. The Musical Mind, the Cognitive Psychology of Music. Oxford Clarindon Press (1985).

103. Spoor F and Zonneveld F. Implications of Early Hominid Labyrinthine Morphology for Evolution of Human Bipedal locomotion. Nature 1994;369:645–8.

104. Standley JM. He effect of music and multimodal stimulation on physiologic and developmental responses of premature infants in neonatal intensive care. Pediatric Nursing 1998;24:532–8.

    The effect of contingent music to increase non-nutritive sucking of premature infants. Pediatric Nursing (2000).

105. Steinke WR. Dissociations among functional subsystems governing melody recognition after right hemisphere damage (2001).

106. Stringer and R. MacKie. African Exodus. The Origins of Modern humanity, New York: Henry Holt (1996).

107. Studert-Kennedy M. The particulate origins of language generocial and cognitive bases. Cambridge University Press (1998).

108. Tolbert Elizabeth. Music and meaning: an Evolutionary story. Psychology of Music 2001;29:84–94.

109. Tallerman Maggy. Did our Ancestors speak a Holistic Protolanguage? Lingua, a special issue on language evolution.

110. Thaut Michel, et al. Auditory Rhythmicity enhances movement and speech motor control in patients with Parkinson's disease. Functional Neurology 1997;16: 163–72.

111. Timbergen N. Derived activities: Their causation, biologic significance, origin and emancipation during evolution, Quarterly Review of Biology 1952;27:1–32.

112. Turk I. Mousterian 'Bone flute' and other finds from, Divje one Cave Site in Slovenia, Ljubljana, Zalozba (1997).

    Wray Alison. Protolanguage as a holistic system for social interaction, language and communication 1998;18: 47–67.

    Holistic utterances in protolanguage: The link from primates to humans in the Evolutionary Emergence of Language: Social Functions and the origin of Linguistic form Cambridge; Cambridge University Press (2000).

Formulaic Language and the Lexicon. Cambridge; Cambridge university Press (2002).

Dual processing in protolanguage. The Transition to language Oxford: Oxford University Press (2002c).

113. Verga Khadem, et al. Praxic and nonverbal cognitive deficits in a large family with a genetically transmitted speech and language disorder. Proceedings of the National Academy of Sciences of the USA 1995;92:930–3.

114. Watson JB. "Psychology as the behaviorist views it." Psychological Review 1913;20:158–77.

"Is thinking merely the action of the language mechanisms"? British Journal of Psychology 1920;11:86–104.

115. Werker JF and Tees RS. Cross Language speech perception: Evidence for Perceptual recognition during the First Year of life. Infant behavior and development 1984; 7:49–6.

116. Wernicke C. Der aphasische Symptpmenkomplex. Bruslau, Cohn and Weigert (1874).

117. Williams HW. Dictionary of the Maori Language. Wellington, New Zealand: AH and AW Reed (1971).

118. Wilson Allen C and Rebecca L Cann. "The Recent African Genesis of Humans Scientific. American (April) 68–73 (a popular account of the genetic evidence from mitochondrial DNA for an African homeland for modern humans) 1992.

119. Wolpoff MH, et al. Conceptual Issues in Modern Human Origins. Research New York Aldine D. Gruyter 28–44.

120. Wray Alison. Protolanguage as a holistic system for social interaction. Language and Communication 1998;18: 47–67.

Holistic utterances in protolanguage: The link from primates to humans. In Knight-Studder-Kennedy and Hurford 2000;285–302.

Formulaic Language and the Lexicon. Cambridge University Press (2000a).

The Transition to Language. Oxford University Press (2000b).

Formulaic Language, Pushing the Boundaries. Oxford University Press (2008).

1. भाषा व भाषाशास्त्र — श्री. ग. गजेंद्रगडकर
2. जनांसाठी जनुके — डॉ. कौमुदी गोडबोले
3. उत्क्रांती — सुमती जोशी
4. मेंदूतील माणूस — डॉ. आनंद जोशी
                     डॉ. सुबोध जावडेकर
5. त्रीकाल् वेध — कुमार केतकर
6. बदलते विध — कुमार केतकर
7. भाषा इतिहास आणि भूगोल — ना. गो. कालेलकर
8. ऐका मानवा तुझी कहाणी — नंदा खरे
9. R.S. Saxena Sumitra Saxena 2008 — Comparative anatomy of vertebrates
10. John Alcock 1993 — Animal behavior
11. Edited By SY Wang 1991 — The emergence of language
12. Philip Lieberman Sheila E Blumstein 1991 — Speech physiology, speech perception, acoustic phonetics
13. Philip Liberman — Eve spoke
14. Steven Mithen 2006 — The singing Neanderthal
15. Scott Brown — Otolaryngology
16. John NE Evane — Pediatric otolaryngology
17. Steven Pinker 1994 — The language instinct
18. Cambridge University Press 1992 — Encyclopedia of human evolution
19. Doreen Kimura 1993 — Neuromotor mechanisms in human communication
20. Mrritt Ruhlen 1994 — The origin of language
21. Lord Russel Brain 1965 — Speech disorders
22. Peter MacNeilage 2008 — The origin of speech
23. Dr VS Ramachandran 2010 — The tell-tale brain

24. Charles Darwin 1859   The origin of species
25. Oliver Sacs   Musicophilia
26. Anthony Storr   Music and the man
27. April MacMahon   Evolutionary linguistics
    Robert MacMahon
28. Christin Kenneally 2013 The first word
29. Noam Chomsky   Nature of language
30. Murray L Barr   The human nervous
    John A Kiernan   system
31. Richard S Snell   Clinical neuroanatomy

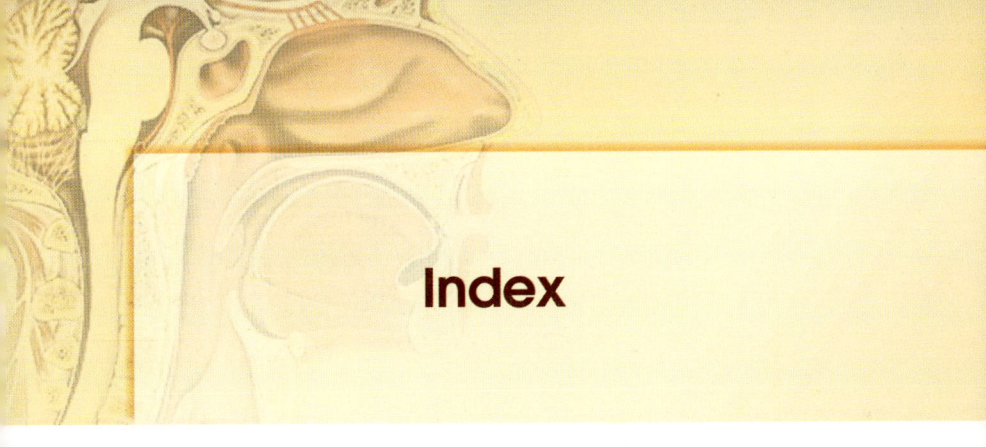

Index